Also by Sandra S. Silver

The - Trust Me On This - Really Good Food Cookbook

Abortion: A Biblical Consideration

Snowtime (a novel)

Pocket Full of Posies

A Cosmos in my Kitchen

The Journal of a Beekeeper

BY

Sandra Sweeny Silver

AuthorHouse™
1663 Liberty Drive, Suite 200
Bloomington, IN 47403
www.authorhouse.com
Phone: 1-800-839-8640

© 2007 Sandra Sweeny Silver. All rights reserved.

No part of this book may be reproduced, stored in a retrieval system, or transmitted by any means without the written permission of the author.

First published by AuthorHouse 11/7/2007

ISBN: 978-1-4343-3021-5 (sc)

Printed in the United States of America
Bloomington, Indiana

This book is printed on acid-free paper.

HOW I GOT MY BEES

Every September my husband Steve and I took our three children to the Danbury State Fair in Connecticut. The older two, Kathy and Blake, were allowed to skip school for this all-day festival into the world of farm animals, feats of oxen strength, lumberjack agility, Ferris wheels and hucksters touting the latest food gadgets and home furnishings. There were always interesting sideshows: the largest lobster ever caught in American waters---a dead monster umpteen feet long housed in its own gypsy caravan. We'd pay the dollar, go into the darkened wagon and marvel at the red crustacean mounted on the wall beside its master who told us three or four unbelievable things then hurried us out the other door.

A man with performing chimpanzees had a small amphitheater near Frontier Village. He was a bored, seedy man whose life had obviously taken an unexpected turn. We suspected him of abusing the chimps and wondered whether we should call the ASPCA.

I always headed for the huge Bingo Hall where regulars played ten cards at a time. I was a once-a-year Bingoist who pawed through the frayed cards looking for a good one. Steve took our little Jesse for rides on the toy airplanes that he both feared and hated as I hopefully placed kernels of corn on B-10 and G-52. I was out-kernelled every time by the pros.

Steve went to the Danbury Fair for two reasons: the food and the food. When we entered the iron gates, he headed for the messy food, the food that tasted good but was a bear to eat: buns with unwieldy sausages smothered in onions and peppers and brothy sauce, hard strips of beef covered with shredded lettuce and a yogurt dill sauce that leaked out over the fingers and onto his shirt.

We all loved the hot fried dough sprinkled with powdered sugar. Usually we ended up eating lunch in one of the huge tents where turkey, gravy, mashed potatoes, pork, sauerkraut and baked ham were served on flimsy paper plates with plastic forks and knives.

For our family the rides were not the big attraction at the state fair. We preferred to sit in the bleachers and watch farmers urge truculent oxen to pull tons of dead weight cement over a finish line. We talked about how the men and animals must have practiced so hard all year for this moment of glory. We wondered at the different ways people choose to lead their lives.

Our favorite attraction, and we had to line up and wait for this one, was the Swedish Sven and his son, Tom, who were experts at sawing wood in half in record time. They had souped-up saws and competed against each other. The father was a giant of a man who won most of the time. The son was big and had youth on his side, but Sven had experience. We'd imagine their lives and how in the world they ever decided to become the Sawing Wonders. They have over the years become part of our family's discussions. They probably would be amazed to know that.

We always ended up watching for an hour or two the man who cut vegetables and fruits with that gadget that worked perfectly for him but wouldn't work for you when you bought it and brought it home. Every year, even after we had bought it and knew it didn't really work that way, every year we hung at the back of his semi-circle and were mesmerized by his skill and spiel.

A woman, the same woman, was always there in the Home Section. She was from Pennsylvania and demonstrated her steel butterfly mold. She'd dip this winged pattern into batter and transfer it to hot oil. Seconds later, it was brought out of the oil and slipped perfectly off the mold and onto a paper towel. She would sprinkle these perfect beauties with some powdered sugar and offer all of us a taste. It was wonderful. For several years I watched her. One year I just had to have that gadget. I bought it. I heated up the oil. I dipped the steel butterfly into the batter. I plopped it into the hot oil. It broke into a hundred sizzling bits of fried batter. The rest clung to the steel mold and later resisted all efforts to remove it. The following year I told her my experience. She said the oil wasn't hot enough and the mold wasn't tempered. Hers were beautiful and delicious, though.

The year before the Danbury Fair closed in 1981 there was a special exhibit that they hadn't had before. It was a small, enclosed room that promised thousands of bees. Nobody but I wanted to go in. The man charged a dollar. I paid my money and entered. It was dark inside. When my eyes became accustomed, I found myself surrounded on all four sides by hundreds of thousands of honeybees behind glass. It was a glass room filled with multiple hives.

At first I was overwhelmed by so many bees and a bit frightened. But I went up to the glass and started looking at them. Millions, maybe, of tiny life forms were there where they couldn't hurt me and I could see them carrying on with their lives. Within fifteen minutes I knew I wanted to have something like this for myself.

I had NEVER been interested in bees. They were always something to stay away from. They could hurt you. So this wasn't an evolutionary interest that culminated in that experience. The experience I had in that little, dark glass room at the Danbury Fair was apocalyptic in the true sense of that word: the "veil was drawn away from" and revealed something to me. I had never thought about the universe of bees.

I stayed in the room about an hour. The man who kept the bees talked to me about them. I asked him if there was any way that I could have something like that for myself, for my own observation. He told me there were observation hives: small but with a glass panel that allowed one to view the life of the bees. He said I would have to find a beekeeper in the neighborhood and he could probably order me an observation hive and get me some bees.

I told Steve that I wanted to do this. He was adamant that I would not. He was very afraid of bees and had once carried a bee that he thought was a killer bee (prematurely arrived in Connecticut) around in a vial in his car for one year. The bee was a victim of one of his raids on a nest outside our front door. He thought it was so big and so mean that someone at the University of Connecticut extension in Stamford should look at it and confirm the fact that it was a killer. I don't know what happened to the bee, but he never had it examined.

Nevertheless, he was not in favor of the idea nor was my son Blake who detested bees, too. He was a partner in the raid on the nest. They had both risked their lives to destroy that nest. But I was going to have my bees. I knew they weren't going to get out and hurt us. The man had assured me it was a very safe hobby.

There wasn't anything under BEES in the telephone directory. But I did see an article in our local newspaper about a man who was going to give a lecture on honeybees at the library in Wilton. I contacted him.

A Cosmos in my Kitchen

That's how Ed W. and the honeybees came into my life.

Ed and his wife, Nita, have a home on a bumpy old road in Wilton, Ct. Nita has some of the nicest flower gardens I've ever seen. "It's the bees," she says. Behind their house they have about 60 hives that stand behind the gardens and stumble almost into the woods. Ed is always busy. He has converted his garage into a honey production center and his brownish thick honey sells in local stores under the label Wilton Gold. He always talks about how he quit the high corporate world to raise honeybees. I told him I wanted an observation hive. He said it could be done. I ordered one for delivery with a package of bees in April of 1982. In preparation for our new experience, Steve and I went to Ed's for a demonstration of hiving. There were about 20 people there. They all were getting outdoor hives. My hiving would be different, would be one that I would have to follow by the book that came with my hive. But Steve and I stood there in the circle that surrounded Ed as he poured out 10,000 bees onto a sheet. We watched them make a beeline into the hive that he had prepared. Steve had tucked his pants into his boots. He had zipped his leather jacket up under his chin and his hands were in his pockets. He thought all the other people were weird types and tried to figure out how I fit into all this.

But we both observed that no one was stung, that the bees seemed intent on entering rather than on stinging. We felt better. He had seen that I would not be deterred and had resigned himself to the situation as one does in a marriage. He was there at the hiving because I had told him he had to help me, that I couldn't hive them by myself. I loved him very much on that cold March morning.

And then I got my bees. And then we all got bees. And then we were all amazed and pleasantly so as one is when someone arrives to stay who turns out to be fascinating and independent and giving.

April 17, 1982

Today at 11:30 I picked up my package of 10,000 bees from Ed in Wilton, Ct. The package is a shoe-box-sized wooden crate with screening on the front and the rear so the 3 lbs. of bees in the "box" can breathe. In addition to the 10,000 (!!!) bees, the package contains a can of sugar water to feed them while they were in transit from Georgia and a mated queen in a tiny cage (c. 3" long and 1" wide) with her wings clipped. She arrives with her wings clipped so that she doesn't in the future lead out a swarm and leave a diminished hive. I put all of them on the back floor of the car and drove home. We hummed all the way!

Amazingly, I felt no fear in handling them. I put them in the cool basement to await the afternoon hiving. I sprayed them with sugar water that both calms and feeds them.

Jesse, age 6, is fascinated by them and totally unafraid. Blake, age 16, was working and vowed he would <u>not</u> be there for the 2:00 hiving! He and most people think I'm "crazy," "weird," "bizarre." The getting and observing of bees in an observation hive is to me a very natural extension of curiosity about life. But I accept the labels and can see what they mean.

I'm putting the whole package of bees in a glass-paneled Observation Hive that will then be attached to one of my kitchen windows. That window has a nice southeast exposure. With an observation hive, the

A Cosmos in my Kitchen

bees can't get in my kitchen, but they can get outside by a little bee gate in the back of the hive. I can watch them from the kitchen-side of the hive through a clear glass window that I can close with a wooden panel. I will not manipulate them for honey, propolis, pollen or wax as other beekeepers do in normal outdoor hives called "supers." I just want to observe them "wild." To let them forage, swarm, exist. Let the queen fail, etc. All the things they do in a natural hive in a tree or bee gum. (A bee gum is any hollow tree or log where bees create a hive.)

It's April---the perfect time to hive bees because they can gather nectar and pollen from the early bloom of the fruit and other trees.

Steve and I started to hive the bees around 3:00 in the afternoon---sunny, 64 degrees. I have read at least thirteen books on honeybees (all I could find), but book learning is different than hands-on learning. Plus I had watched Ed hive five packages of bees. Daunted, I proceeded.

I first sprayed the cage thoroughly with sugar water (3 cups sugar and 1 1/2 cups water.) That calmed the furor of 10,000 little bees looking out at me and buzzing and pawing and wanting to "get" me. I banged the cage on the ground. Most of the bees fell to the bottom of the cage. Then I took my all-purpose "hive tool," a 10" long standard metal tool with a scraper on one end and a hook on the other for extracting nails. With it I pried open the board on top, popped out their feeding can and the queen cage and put the board over the cage so more bees couldn't escape. A bunch, maybe 40, escaped with the extraction of the sugar water can and the queen cage, but they were docile and solely concerned with the whereabouts of the queen and the other bees.

We, Steve and I, had on bee veils and used our bare hands. Now, at this juncture, I made a strategic error because of haste, lack of hands-on experience and over-confidence born of fear. I opened the WRONG END of the queen cage! I thought that one end had a cork and candy plug and the other end was solid wood. Wrong! Both ends have cork plugs. The one I opened had a cork plug and no candy plug, so the queen's attendants (those female bees enclosed with her in her cage) immediately began to exit the hole. I had thrown the little plug on the ground and couldn't find it.

Steve became very upset! "You blew it! You blew it!" he yelled. Not wanting the queen (who in a package of bees has her wings clipped so she can't fly) to escape, I dropped the queen and the cage into the top of the hive and clamped on the lid. At least I knew where she was even though her attendants were now taking wing from the grass under my feet.

I was trembling. Steve was verbalizing. Bees were buzzing. Blake (he did come for the hiving) and Jesse were in our car with the windows rolled up looking horrified. Even though they didn't know what had happened, they knew something was awry from Steve's reactions. Friends, the H.'s, who had driven over for the novelty of a hiving were also in their car with the windows rolled up. I looked at them all. They were appalled even though they didn't know WHY they were appalled.

I took off the too big bee veil, went into the kitchen, lit a cigarette and called Ed, "I made a terrible mistake. I opened the wrong end of the queen cage. The queen's now in the upper chamber of the hive. How do I get her out, back in her cage and into the lower brood combs where she's supposed to be?"

Ed drew a deep breath. I'm sure this was the fourth or fifth call this day, this day when he had handed out hundreds of packages of bees for outdoor hives to dozens of people.
"Okay, calm down. Just get in there, pick her up and coax her back into her cage right there on your kitchen table. Is she all right?"
"Yes, I think so."
I prayed and went back out.
Steve said in a loud voice, "You're not doing another thing, Sandy! I'll do the rest. What did Ed say we had to do?"
"He said to pick her up by her head or thorax, not by her abdomen and put her back in the queen cage."
"Pick her up!"
"Yes, Steve. I'll do it."
"No, no! I'll do it!"

I had, of course, lost all competency because I had "blown" it. I was actually amazed that he who had so opposed my getting bees was prepared to do the most hands-on of all maneuvers---picking up the queen with his thumb and forefinger. I was not anxious to touch Her Highness and rather in awe of his male machismo.

Meanwhile, the H.'s slowly drove away, waving and forcing smiles. They knew the hiving had become a family feud and did not want to observe Steve and me so "vulnerable." Blake and Jesse were mouthing from the safety of the car, "What's the matter?" I smiled at them. "It'll be all right."

We took out the three combs at the top of the hive. There she was with the white dot on her back. The people who sell bees always designate the queen with a mark. She was on the queen excluder (a series of metal bars designed to keep the queen in the bottom brood comb and out of the upper chamber honeycombs). Steve grabbed the queen cage and stuck his two fingers in to get her.

"Where the hell is her thorax?" he yelled.
"Above her abdomen, Steve."
Disgusted, he pushed his thumb and forefinger closer and closer to her.
She gave a warning buzz.
Steve made a hasty withdrawal.

I went in and took her gently by her thorax. I had prayed in the kitchen that the Lord would help us. She buzzed and flapped and quivered in my fingers. I showed her the hole to her cage and she went gratefully in. I clamped on the plug, inserted her cage into the lower right hand side of the hive and closed it up. Done!

Now to hive the 10,000 little ones. I smacked them down into the bottom of the cage again, inverted the cage over the matching hole on top of my hive, removed the cardboard cover and they started pouring into the hive. It was like birthing a baby in the sense of anxiety and exhilaration. I guess only those who have hived bees can understand this poor analogy, but it was a very similar feeling. This hiving was a birthing with a potentially lethal complication as was my emergency Caesarean delivery of Jesse. (Julius Caesar was delivered that way according to Pliny the Elder and that's why the operation is named after him. Pliny the Elder distinguished from his nephew Pliny the Younger died in the eruption of Mt. Vesuvius in 79 A.D.)

We tapped the cage again, put a rock on top of it all and then Steve and I laughed. The boys got out of the car. By this time no one was afraid of them. We all rejoiced and watched them flow into the clean, herb-scented hive. I had rubbed their new home with fresh thyme and rosemary to greet them.

Steve and I went to the movies that night. When we got home, it was raining like crazy. The water was almost up to the bee gate. Dear Steve intrepidly picked up the whole hive and carried it onto the back porch. All but several hundred had left the bee cage and entered the hive, so we smacked most of them into the hive and closed it up. Steve carried it into the kitchen and installed it in the window next to our kitchen table. The bees were very lethargic. They were syrupy, wet and disoriented so there was no problem.

I watched them from 11:00 that night to 4:00 in the morning through a magnifying glass.

April 18, 1982

The bees are all over the side of the house exploring their new environment! The hum from the hive is intense and gives a great feel to the kitchen. I'm going to love these bees!

They are all over the brood comb that is in the bottom of the hive. The honeycombs on top of the hive are almost totally ignored except for transportation purposes. It's sunny and in the 70's. The Scilla siberica (about 10,000 bulbs---one for each bee) are the main attraction outside. Inside they are busy trying to eat the sugar plug away from the queen to get to her. By now they know her scent and have accepted her. (If they don't know a queen's scent, they will kill her. That's why she has to be isolated from them in her own cage. They're very territorial.) Twenty-four hours a day---there is never a time without bees eating away at the plug trying to get their queen out. All the bees on the bottom brood comb are concentrated right above the queen cage. They are never far from where <u>she</u> is.

I can see tiny bits of pollen that they have deposited on the honeycombs. We all watch them constantly!

April 19, 1982

Sunny and they're out in force. The noise in the hive is intense---busy as bees. It's amazing. The mass of bees right above the queen cage has been laying wax comb. Perfect, snow-white, ROUND combs WITHIN HEXAGONAL FRAMEWORKS. The HOLES are not eight-sided. They are round. Only the super-structure is hexagonal. What I mean is: the little wax combs start out eight-sided. Then they work on them to make the actual hole as smoothly round as possible.

First they build the hexagons with the wax secreted from the wax glands in their abdomens. Then they, with their little heads inside the hexagon, begin to work the hexagons to make them round and smooth. Nature loves smoothness, curves, softness---not jagged edges. The jaggedness of Nature is only the skeleton around which the real life hangs. The bare lineaments of a winter tree and then the smooth, puffy, soft ball in summer. Our bony, bare skeletons and then the smooth curves of flesh. But both---the jagged and the smooth---are necessary. The skeleton is inert, while the soft smoothness is life, living.

The hexagon of the honeycomb is really a circle within an eight-sided superstructure. Start with a skeletal hexagon; round out the hexagon with wax; end up with a circle inside an eight-sided structure. I'm reminded of the hexagonal quilt pattern. That's where "quilting bee" came from. The women made hexagons and were as busy as bees as they quilted those lovely comforters. They didn't round the hexagons as the bees do, but the quilts covered the softness of smooth flesh.

This afternoon around 4:00 I think the queen came out of her cage. The bees ate through the candy plug and she emerged. I have not seen her yet. I'm sure they hurried her away from all these peering faces looking into the glass and have taken her into the darkness behind the brood combs.

I have three racks of combs in this little observation hive. Each rack has two sides, so I have six areas where they can store honey (on top); lay brood and store pollen (on the bottom).

Two reasons I think the queen is out:

1. At 4:00 P.M. I looked at them and there was a massive density of bees on the comb right above her cage---crowded, faces in, abdomens out into a focal center point looking like a complex flower, a cactus dahlia. Furious excitement and activity in that circle. I just felt the queen was under there someplace.
2. I have been constantly observing the entrance to the queen cage. One or two bees would be in there at the same time. Now I see, I count up to four bees exiting from her cage hole. That many could not be in there at the same time if the candy plug was still intact. There is a lot of activity around her cage, though. So, who knows?

Saw the first drones. Was wondering where they were. Huge guys on the upper honeycombs only, squeezing in and out of places, being fed orally by the female worker bees. I know the drones don't work, but feel they must have other reasons for existence other than mating the queen and the morale of the hive.

Since I've had my bees, I've reflected on people saying: "Those poor drones out there." "All the poor drones trudging to work." This use of the word "drone" implies that a drone is an undervalued employee working himself to death in obscurity. Where in the world did this usage of that word come from?! DRONES DON'T DO ANY WORK. In our language we imply they do ALL the grunt work. Etymology. Ever since I took Latin and Greek, I LOVE the study of word origins. A word comes into a language and over the centuries of use many things can happen to it. It can MAINTAIN its original meaning, can be ELEVATED to a better place or can be LOWERED from its original meaning. "Drone" from the Old English word "draen" came into our language meaning "idle, lazy worker." "Drone" with the common usage of "unappreciated worker" has definitely been ELEVATED in meaning. (In the bee world, however, it still retains its original meaning of "lazy, idle.") When "cute" came into our language, it meant "bow-legged." I guess there were lots of cute girls who were bow-legged because that word has, also, been elevated. I have a "queen" in my hive. "Queen" comes from the Old English word "cwen" that meant "wife" or "honored woman." That word has been elevated to a title only one woman can have. "Villain" came into the language meaning just "farmhand." This word has undergone PEJORATION. It has been lowered in meaning, has developed negative connotations. Maybe too many farm workers were untrustworthy? "Manure" originally meant "worked with the hands." The word has definitely undergone pejoration. (If someone used the word "shit" around my grandmother Meme, she would say, "You just had something in your mouth I wouldn't have in my hands." I loved that colorful reprimand!) There are some words that can't be traced back to their original meanings. "Bee" is one of those words! "Honey," however, has a long and traceable history. Old English "hunig." Middle English "honi." Today "honey." (There is a town in the Hyndburn district of England called Huncoat. In O.E. "hunig" = "honey." "Cot" = "storage." So Huncoat was the place where they stored their stashes of honey a thousand years ago.) "Nectar" comes from two Greek words "nek" meaning "death" and "tar" meaning "overcoming." Nectar was the drink of the gods in mythology. Only the gods had overcome death and were immortal. "Pollen" comes from the Latin word "pollen" meaning "dust, fine flour." It has retained its original meaning. I know "wax" comes from O.E. "weax," but I can't find its original meaning. "Hive" comes from O.E. word "hyf" that meant "hull of a ship." Speculating: maybe on long ocean voyages in the 7th century they kept the precious honey stored in the hull of the ship. Or maybe they kept bees in hives shaped like the hull of a ship. Or maybe it's just the "container" idea. There are "workers" in my hive. "Work" comes from the German word "werk." Can't find the original MEANING. I'll keep trying. All the time people say "entomology" (study of insects) when they mean "etymology" (study of word origins). That little "n" makes a quantum world of difference.

April 20, 1982

Still haven't seen the "queen" (a word ELEVATED from meaning all "wives" to meaning just "one royal woman"). I still think she's out and around though. Activity around her cage has diminished greatly, but some bees still go in and out of it. It's as though:

1. They still want to check out her home, i.e. where SHE was---a sort of tourist attraction.
2. They are making sure nothing else is there.

3. They go in to get her smell that still lingers.
4. Just random exploration.

The wax comb is so beautiful. I love the way they laid it right over her cage---ready and waiting comb for her eggs---right where she would exit and see how busy and eager they were for her to perform her sole raison d'etre. I think the comb is probably laid in that part of the hive on all the other five sides that I can't see. So logical.

"Comb" comes from the O.E. word "camb" meaning "toothed object." Now when or why they started to call the bee's honeycomb "comb," I'll never know! I'm looking at the comb. I guess you could stretch and say the ridges of the hexagons are comb-like, but that's not a satisfactory explanation. Everyone thinks of a comb in relation to hair. "Comb" is a "toothed object." That I get. But other meanings of the word "comb" are hard to figure out. A crest of a wave is called a comb. "Beachcombers" are both rolling waves and people who live, usually vagrantly, by the beach. "Comb" also means "to search everywhere." She combed the house with a fine tooth comb. THIS IS REALLY A COMPLICATED WORD!!

"To draw comb," as laying wax is called, they hang in beautiful necklaces. First, one bee holds on to the top of the wood of the comb with five of her six legs. That leaves one back leg free. The next bee grasps that back leg with one or two of her middle legs. She folds down her two front legs. Her back legs are free. The next one grasps her back legs up near the front of the leg.

1. The next one holds onto the previous one's back legs up near the top of the leg.
2. She holds on with her middle legs.

3. She folds down her front legs.
4. That leaves her back legs free to repeat the cycle.

Some necklaces (that's what bees hanging and laying comb are called) go from the top to the bottom of the comb with sixty (60) or more bees in three or four strands. Some are single strands with just eight bees hanging. All are rather inactive to view, but are suspended, hanging, purposeful. All the action is happening under the necklace with each bee extracting from her wax glands tiny particles of wax and affixing the wax to the superstructure of the comb. When they are laying comb, they really do look like unclasped necklaces.

The comb is the big attraction now. Like magic, it takes form and presence. They are now working on the whole comb except the corners. I guess they leave the sides and corners until last.

They lay the wax comb in sections up to down---not side to side. Vertically not horizontally. They laid this bottom comb first---top to bottom on the right hand side of the observation window over the queen cage. Then they moved to the opposite left side of the hive and laid that comb---leaving the middle free. They started moving toward the middle from the right hand side. The comb is wavy, hilly to appearance. Now, the bottom brood comb has some wax comb, except for the ends and sides.

The comb-laying priorities on the only side I can observe seem to be:

1. On the right over the queen cage.
2. On the left hand side of the brood comb.

3. They work to hook up in the middle in a rather random, hilly fashion.
4. Do the sides and the corners last. Probably for transportation reasons.

Still no drones in the work area. There's a dead one at the bottom of the hive. Each day, each night I watch them remove their dead ones and put them outside the bee entrance (gate) on the windowsill.

Also, I watch them clear away feces. They put it outside the bee gate, tuck it in between the wood and the glass and, amazingly, even shove it out the tiny air screens on either side of the hive where it falls onto my kitchen floor! The hive is really clean considering so many occupants in such a small city!

They are currently taking three bottles of syrup water every 24 hours. That's one of the ways they are getting sustenance for all the comb building. Comb laying requires a lot of nectar consumption. The sugar water is a compensation for that.

Looked outside. It's late. Cloudy. But I saw some stars! Imagine. Utah has an official astronomical symbol! That's strange. I knew that states had State Flowers, State Trees, State Mottos, etc. But State Astronomical Symbols?! Utah's astronomical symbol is called THE BEEHIVE CLUSTER! I knew that Utah's nickname was The Beehive State because Brigham Young kept bees in baskets called skeps. That's probably why this Mormon state picked that "astronomical symbol." My bees are so ingrained all over the earth and now all over the heavens. Just looked up the Beehive Cluster. It's impressive. It consists of over 200 stars and is about 600 light years away from us. The scientist calculated that if I wanted to be propelled to the moon at the speed of 186,000 miles per SECOND (the speed of light in a vacuum), it would take only 45.1 seconds for me to arrive there! Not bad. But if I wanted to go at that speed to Andromeda, our CLOSEST constellation, it would take 88.1 MILLION YEARS! Those are LIGHT YEARS. SO YOU'D HAVE TO MULTIPLY 5,865,690,000,000 (A LIGHT YEAR OF 365 DAYS) BY 88.1 MILLION YEARS TO KNOW.... There you go. It's the INFINITY of it all that boggles me. My bees are an entire universe. My body is an entire universe. Every atom in me is an entire universe. And then there's the actual universe of light years and the relativity of time and the curvature of space and....By the way, if space is curved and the universe is ovate, what's on the OTHER SIDE of the curve? Our universe is a CLOSED STRUCTURE! What's BEYOND THE CURVE OF SPACE? I love Einstein's quote: "God does not shoot dice with the universe." God's the Great Mathematician, He. Love William Blake's watercolor of The Ancient Of Days. He depicts Him as bending down and measuring the universe. Sad that I'll never understand the math of the universe even on an elementary level. But I do get the AWE part!!

April 21, 1982

Raining, cold. No bees outside. One bee wandered out for about 20 seconds and then flew in again. The lovely hum is not present. You can't hear a thing inside the hive. They are busy, but slow, mute, laying wax. No excitement about anything apparently.

12:00 noon. Just now I saw a shaft of sun shoot across the trees. Almost immediately, I mean IMMEDIATELY, I heard noises inside the hive!

The sun. Maybe some of their noise is geared toward the sun? Sun means:

1. They can go out.
2. They can get nectar.
3. They can get pollen.
4. They can get propolis.
5. They can "play."

I have observed that they have made two holes, perfectly round, through the comb to the other side of the hive. The holes are made in the lower right hand side of the comb. These two egresses will relieve the congestion and traffic problems in the hive.

There are 10,000 bees in this hive and there are only six (6) combs. This package of bees that I got from Ed from Georgia is intended for a regular outdoor hive that has at least twelve (12) combs. In a "normal" hive there are 50,000 to 80,000 bees. My bees will swarm for sure! So many, so little space. But the two round holes have relieved some of the traffic problems allowing them a short cut between this side and the other sides.

It's 12:15 midnight. I'm observing. Now they have made a long hole on the left hand side of the comb. It's about 3" long and 1" wide. They are passing in and out of it. Seems that comb-laying is a rather fluid thing. Comb laid; comb chewed through as needed.

A very quiet hive. There is no hum; no movement on the floor of the hive; absolutely minimal movement in the whole hive---4-5 bees are running around. The rest are clustered with marginal movement on the brood comb. I hope the queen is all right. I hope this is not a dispirited hive because the queen is dead. They are certainly quieter than they were the last four nights. If she's okay, maybe this is just real intense work time or rest after intense excitement and work? I won't know the answers until I see the queen.

Funny, how everything revolves in essence around the queen. Yet she's a captive monarch ruling absolutely at the discretion of her subjects.

Even the drones are out tonight lolling on the honeycomb. The other nights they were not observable. They are quiet, too.

Other bees are still working on the smoothing of the inside of the combs with their heads. But I see no necklaces drawing comb. Saw some this afternoon though.

They have perforated the comb where the wires are---the wires that hold the superstructure of the comb. Maybe that's the weakest point for them to go through.

For the first time I see no dead bees on the floor of the hive.

This afternoon I saw a bee parade a dead bee all over the hive and comb. She carried the dead bee with her proboscis hooked on to the dead bee's proboscis. Proboscis to proboscis. Mouth to mouth. Imagine carrying your own weight very spritely all around your yard for an hour or two! Carrying someone your own weight and size BY YOUR TEETH!

April 22, 1982

9:00 A.M. Hive still quiet. I'm worried. It's sunny, 52 degrees. No bees out.

9:31 A.M. 55 degrees. One drone flew out. Some buzzing in the hive. During the night, they consumed c. 5" of sugar water vs. their usual whole bottle.

9:40 57 degrees. Little more activity in and out of the hive. The drones are going out. Interesting: when they take off from the bee gate, they go straight out through the open storm window into the air. The workers, however, often drop to the windowsill, fly upwards and wander around on the outside of the window for a while before they finally "find" their way out. Perhaps the drones with their larger eyes see so much better that it is a straight horizontal take-off.

The females frequently get "trapped" between the window and the storm window until they finally fly out. Just a thought.

The foragers LOVE the carpet of thousands of little blue scilla under the big, 300 year old copper beech. I can see several now returning with the blue pollen packed into the little baskets there on the sides of their hind legs. The other day I lay in the bed of scilla with my magnifying glass and watched each bee load up on the beautiful blue pollen. They just packed in the blue seed, tucking and tamping down the pollen. They haven't deposited any of it on this side of the comb though.

When I was a little girl, people used to say if something was amazing: "Well, isn't that the bee's knees!" I think they were talking not only about how wondrously made the honeybee is but about the actual knees of the bees. They have three sets of jointed legs. The hind legs are where they have their pollen baskets. The two baskets are located right at the joint of the legs (hence the bee's knees). When she if harvesting the pollen from flowers like the tiny scilla siberica here, it seems she pushes the blue pollen into her third leg by

A Cosmos in my Kitchen

the other two legs. I've read that in order to pack in the powdery pollen, she will regurgitate nectar she has ingested and pass it along from her mandible to the other legs that in turn place the nectar ON the pollen making it hard and not so powdery. She normally loads up with about 5 ounces of packed pollen before she flies home with her back legs dangling in the air, heavy with pollen.

She only weighs about .0004 ounces and is a mere $3/5^{th}$ of an inch long. So she's flying back to the hive with pollen packs that are over five times her weight. Take your weight and times it by five and then carry that weight WITH YOUR LEGS for 1-3 miles. Little ones, you are so amazing!

6:15 P.M. Hive quiet. Floor of hive is immaculate. No bees are using it at all. They are still all at work on the comb, I trust, because I can only see one sixth of the hive. But they're not excited.

On the left hand side of the hive, they have begun to fill in the space between the comb and the outside of the hive with wax. Maybe this is to root it firmly for the load of brood, honey, pollen. A little reinforcement masonry?

On this left side there is not one hexagon that hasn't been filled in with wax. Most hexagons are high and seem finished, but around the edges of some only three sides have several coats of wax. They need to be finished. I wonder if all the hilliness goes away and the comb becomes flush, so to speak? I am worried about the low level of activity. The queen?

12:00 midnight---definitely something wrong. NO activity. NO hum. A phalanx of bees is on the glass looking outward at me. Another "military formation" is on the comb facing away from me. They are taking almost NO sugar water. All huddled together---silent, dispirited. The queen is probably dead. I, no doubt, hurt her when I picked her up. But what do I know? Maybe this is normal for a six-day old hive, but they are different in the last two days.

April 23, 1982

Hallelujah! My bees are okay! Today it was in the 60's and sunny. They were out in force. The hive was busier than it ever was. Called Ed and he said they were just too cold yesterday to do much. The wind chill factor yesterday put the temperature in the 40's. He said to check and see if they are bringing in pollen and to keep checking for brood on the combs.

I took my magnifying glass outside and lay in the scilla again to watch them load up on pollen. They are tucking the blue pollen into the pollen baskets on each side of their back legs as fast as they can. So maybe she is alive and well.

They're still polishing up the combs individually--- filling in little by little the bottom, sides and edges of the hexagons. The passageways that they have created through the comb have increased in numbers. (The foundation of the comb is a structure of wires on which they lay their wax comb.) They have now chewed through four of the foundation wires. The holes to pass through are made at the wires rather than in the middle between the wires. They've chewed through eight or more, each hole several inches long only on the bottom of the comb. They pass to and fro squeezing around both sides of the wire.

There is a lot more activity and attention on the bottom brood combs now, but so far no wax laid or honey deposited on the upper combs. Just reconnoitering.

They tuck their feces between the glass on my observation side and the wood that separates the honey and brood comb. They use their proboscises for this work. The proboscis (mouth, feeding tube) is LONG for their relative body size, I think, and pointed at the tip. It's a pretty red-pink. A useful tool to have.

12:30 A.M. Just watched a fascinating thing. A bee was going to "rest" on the glass facing me. She was belly to me, back to the comb, so I could observe her underside very well. She had to go through a lot to get her "grip" on the glass. First, she rubbed her back legs against the sides of her abdomen and kept doing that until they stuck to the glass. Maybe she was getting something from her abdomen, an adhesive, or maybe it was for a good "feel" (clean) like we do when we wipe off our palms on our pants or shirt when we need a good grip. Then she rubbed one of her middle legs between the two back legs to get the same thing on them or "clean" them. Then she redid her back legs on her abdomen, felt the grip and then did her two front legs with the back legs. They do have six little legs. When she was all set, she "rested"---firm, secure, unmovable. You can see the folded or tucked proboscis when they rest belly up like this. It's the only RED on them---bright red between shiny black. Beautiful.

The sugar water consumed has dropped from three bottles every 24 hours to about 1 and a half bottles per 24 hrs. That could mean:

1. Most of the comb is laid.
2. More outside nectar is being brought into the hive and they don't need the supplement.
3. All of the above plus more that I don't know

The drones that I can see all "sleep" on the wood in the space under the honeycomb at the top of the hive. I NEVER see any in the bottom of the hive so far. That's the place where all the work is going on and they don't work. They sleep head into the interior of the hive, rounded bottoms out, all lined up sideways like parked planes.

April 24, 1982

9:05 A.M. They are so busy and excited this morning that they have fogged over part of the glass in the bottom of the hive. Little beads of condensation all over the inside.

At this point in the hive's development, we can see through the hexagons on this side to the hexagons on the other hidden side, because the ones over here are still empty. Blake noticed that some of the hexagons on the other side were filled with something---here and there something dark and something golden. Maybe the dark ones are the blue scilla pollen and the light ones are developing larvae?

6:30 P.M. Blake said, "Have you seen the honey?!" I hadn't looked at the hive since this morning. All day they were very busy. Behold, they have laid in an incredible amount of nectar on the lower brood comb! I'd say one third of the comb is nectar-filled. When the light hits the comb, it glistens like crystal. It's gorgeous! I would have thought they would place the nectar in the top honeycombs, but they haven't even laid comb up there yet. I wonder if there will be brood on this side of the bottom comb or if they will just use it for nectar storage, especially since we are always looking in on them.

This was by far the nicest, warmest day we've had this spring and my bees were out from early morning until just about half an hour ago. Some few stragglers are just coming in now, nectar crops full, pollen baskets bulging.

They've only taken three quarters of a bottle of sugar water. Who needs that synthetic stuff when you're drinking the real thing! Just imagine that this little, teeny bee is the ONLY way the sugar hidden in all God's fields of flowers and trees can be gathered and turned into man's oldest sweet---honey! Man has only "domesticated" two insects for his use---the honeybee and the silk worm. Must read up some day on the silk worm and the fabled Silk Road that brought those beautiful cloths woven from the woof of those little worms from China to the rest of the world.

Note inserted later: Just read up on the Silk Road. Fascinating. I have a map of the routes of the Silk Road before me now. I say "routes" because there wasn't just one well-worn road from China to the Mediterranean. I'm looking at a northern route, a southern route and a middle route. All the routes started in China on the Pacific Ocean and wound west through deserts, mountains and icy passes. 5,000 miles of cold, snow, ice, blistering desert heat, impenetrable mountains and bandits. As the Road expanded over the centuries, it eventually ended in cities on the Mediterranean Sea.

There WAS a "first person to travel the Silk Road." In fact, he pioneered the Silk Road. His name was Chang Ch'ien who commanded the guards in the palace of a Han emperor, Wu-ti. In 138 B.C. Wu-ti sent Chang Ch'ien and 100 men on a mission west to seek an alliance with other tribes. He was captured by one of the tribes and married one of their women. But he was the proud and loyal ambassador of a king. He escaped and continued to journey west on what became the northern route. When he reached the location of one of the tribes, they had migrated. After many vicissitudes, Chang Ch'ien pioneered the southern route back to China. By the time he reached home 13 years later, only one of the original hundred men were with him. But he did bring back copious, first-hand knowledge about the geography of China and the traditions and practices of all the tribes in the west of China. Plus, he told his emperor that there were peoples even farther to the west: the Arabs, the Romans and the Parthians. He is honored in Chinese history as the man who opened up the Silk Road.

It was from the barbaric Parthians that civilized Romans first learned of SILK. In 53 B.C. the Roman armies were fighting the Parthians. Suddenly, the Parthians unfurled gorgeous banners that shone in the sun. This new material caused the Romans to abandon the fight and go home! What was this new material and where did those barbarians get it? They learned it came from the Seres, "the silk people," far to the east. Rome wanted that new shiny material. Ironically, the Parthians became the silk broker between China and Rome. Every Roman wanted silk. It was literally worth its weight in gold. Even though the treacherous Silk Road had been in use for generations, Rome's huge market for the material made it a well-worn highway. What propelled the Chinese to start out on such a many years' long and peril-fraught journey? Money. The entrepreneurs loaded up their caravans with silk, jade, gold, tea and spices. These were the high-price and easy-transport commodities that the people in western China, Arabia and, most of all, Rome craved. Gradually trade went both ways. Caravans carrying gold, bronze statues and ornaments, precious stones and glass left the west on the Silk Road east TO China. The Chinese didn't have glass. The demand for it from the west was as intense as the demand for silk from the east. The Silk Road is at least as famous as another ancient road, the Via Appia in Rome. I like that China/Rome link. The Inscrutable East and the Scrutable West.

April 25, 1982

Glistening nectar all over the comb! Almost 80 degrees today. They are out in force gathering nectar and pollen. I'm sure the orange/yellow filled hexagons on the hidden comb are pollen or brood. Still no sign of the queen on this side.

There is a pattern, somewhat, to the way they are filling the comb with nectar. The whole top of the brood comb is nectar-filled in an in-and-out curvy manner.

They are still putting wax between the comb and the wood to make it all one piece. Still, also, making wax on the outsides of the comb on the wood.

I hope the workers, who see us all the time, will allow the queen to lay brood on this side so I can observe the entire process. Even though the nectar takes up the whole top half of this comb, there is still room for brood on the bottom half. Maybe they intend to use this whole comb for nectar and keep her and the brood away from our prying eyes. I hope not.

April 27, 1982

The queen is laying on this side!! I've seen the queen!

4:45--5:15 P.M. She is laying in the middle bottom left of the brood comb. Thirty seconds---one egg; 17 seconds---one egg. Now her attendants are grooming and feeding her. 17 seconds---egg; 20 seconds---egg;

19 seconds---egg; 19 seconds---egg; 23 seconds---egg; 42 seconds---egg. She rests for four (4) minutes. 17 seconds---egg. Now for several minutes she moves slowly up to the top of the comb where nectar is stored followed by her circle of midwives. Maybe she's sipping nectar or checking the stores to make sure they're doing their job as she does hers or maybe she's just taking a rest from laying. Maybe one of or all of the above or something I can't fathom. After her first series of layings before I got the stopwatch I now have, she also took a tour up to the nectar stores.

She's back in her laying area and is in the hexagon for 44 seconds---egg; 21 seconds---egg; 37 seconds---egg; 17 seconds---egg; 35 seconds--egg; 1 minute and 47 seconds---egg!; 13 seconds---egg. Now she's walking over to the right of the hive and she's in a hexagon. 15 seconds---egg; 14 seconds---egg. Boy, I'd hate to pop out all those eggs day after day, month after month, year after year for up to 5 years! With all those "encouragers" around me all the time egging me on, to make a pun.

Now she's ambling through the hole to the other side of the comb where I can't see her. They'd widened this hole between the wood and the comb. Maybe they widened it today in preparation for her coming through? So maybe they knew AHEAD of time where she was going to lay today?! Interesting that she laid the last few eggs on the right side of the brood comb right before she exited.

Also, she seems to feel out with her head or some part of her head the hexagon she will lay the egg in. Then she proceeds forward and lays the egg with her abdomen. When laying, her head is several hexagons away from where her abdomen is inserting the egg. As she ambles forward, some hexagon must appeal to her, eye-head-wise. When she is several hexagons

A Cosmos in my Kitchen

away from the designated hole, she knows it is time to insert her abdomen in the cell that she cannot now see. Does that make sense?

QUEEN EGG-LAYING CALCULATIONS / RATHER APPROXIMATIONS

1. She averaged, if you calculate 18 eggs in a 30 minute period, c. 1.7 minutes per egg.
2. She laid 18 eggs in 30 minutes.
3. She roamed twice around the top of the comb.
4. She rested near brood cells for 4 minutes one time.
5. There were 12-18 bees in attendance in a rough circle around her at all times. They tended to her abdomen, stroking it with their legs. She, also, occasionally wiped her back legs on her abdomen before she would lay---maybe to get a grip.
6. Except for her attendants, all the other bees busied themselves as though she wasn't there. I did notice, however, that when she moved to the top of the hive in her roaming, the bees made space ahead of and around her.
7. The last three eggs were quickies, 13, 15, 14 seconds respectively. She was perhaps hurrying to get it over with? Perhaps she can control how long it takes her to lay an egg? Or maybe it is like human delivery. You never know how long it will take to give birth?

7:40 P.M. I just watched her lay 17 eggs. She laid each one fast, couldn't have taken more than 20 seconds per egg.

She either senses with her abdomen where she will lay the egg or the attendants instruct her. She definitely lays! I see her court of ladies in waiting even push her along as she moves. She sometimes loses her balance, like a tipsy one.

This time she's led around. She's their prisoner, hemmed in, egged on. Now she appears as one drugged.

Yes, they lead her and tell her! I just saw a bee drag the queen's head forward and then stop her. She then plopped her abdomen down. They even aimed her! Yes! Now they're pushing her to the right. When she lays, it's like going for a hard BM. She props her back legs against the comb and bears down. The other legs rest on adjoining hexagons. She's still laying in the left middle, lower part of comb. They pull and push her forward. She appears dazed and loses her balance a lot! She's the queen of the hive, but she pays a heavy price for her crown!

8:00 P.M. After twenty minutes she's crawled down to the left corner of the hive and has gone to the other side between the comb and the wood in the place they made for her. They had, also, widened the other side for her to enter and exit.

April 30, 1982

Blake and I observed the queen for about 30 minutes. This time we watched her on the extreme lower left of the brood comb. She put her head over the hexagon as if inspecting to see if there was already an egg in there. When she saw it was empty, she would move forward two or three hexagons and insert her abdomen, swiftly, deftly. Today none of the drugged, dragged around manifestations of the previous observation. She's recuperated. Average laying time---20 seconds per egg.

She was much more in control today, less tired than before, more decisive, in control of her laying and her courtiers. Perhaps she decides where to lay AND they direct her?

Her pattern on this side of the comb seems to be so far that she lays for c. 30 minutes and then goes to the interior of the hive. Then she returns about 10 minutes later. Is she resting over there or is she laying?

Just watched a worker bee trying to get a dead drone out from the top honeycomb. Over and over she carried him in her mouth by his penis. I'm almost positive it is the penis because they don't have stingers and it comes out of the middle of the abdomen at the end of the abdomen.

Also, by happenstance---isn't it all by happenstance?---I saw a drone die. He just fell like a shooting star right before my eyes. He landed between the glass and the comb in the top part of the hive. When he hit, that long, ragged, white thing popped out of his abdomen. Immediately, a worker grabbed it and started to haul him away.

1. If it is the penis, that means every female worker has a chance for some kind of sexual contact---however futile.
2. If it is the penis and it ejects at death, it is a provision for removal from the hive. The worker is much smaller than the burly drone and this worker seemed to have no trouble dragging him along by his penis. It would be much harder for her to get a firm grip on his head, legs, thorax or abdomen. Interesting speculation, anyway.
3. I really feel it is the penis that ejected at death. Also, the drone's penis must be EXTREMELY well-attached to his body since only the force of orgasm

with the queen can dislodge it (and, of course, in that instance part of his abdomen blows off as well!).

I have, also, observed female bees, unable to drag the drone through the opening between the bottom of the upper comb and the queen excluder, dismember the drone and take out his head, then thorax, then abdomen. His thorax is so round and furry that it often catches on the wood. The head and abdomen have little trouble getting through, but it's hard work to get the thorax out.

May 2, 1982

Steve and I have gone white-water rafting with Noelle and Paul L. in West Virginia. I, of course, had no intention of actually white-water rafting, but I went along anyway. Yesterday I dropped the three of them off to go rafting. I was driving along a bumpy W. Va. back road when I saw a bunch of outdoor hives stumbling down the hill. There was a sign in front of the clapboard house that said HONEY. I stopped. Thirteen-year old Larry L. took care of the 60 hives. He and I ended up spending several hours together. He had purchased two hives when he was only eight years old and has steadily increased his stock. He gets an average of 50 lbs. of honey per hive. Some of his hives even yield 100 lbs. a season! He said the honey is mainly clover and poplar honey.

Each year he gets several new queens ($5.00 a piece, clipped and marked). That way he is assured of a continual good crop of bees. Last year he said he missed a gyne cell (a large protruding cell which contains a developing queen) that the bees were hiding in his hive. He had clipped off all the other developing queen cells and just didn't see that particular queen cell. So when he introduced the new, clipped queen he had bought, the bees killed her right after they had chewed away the candy plug. They had already established their own queen. Larry was furious.

The bees are always hedging their bets. They can never be sure that something won't happen to their current queen, so they build new queen cells (aka supercedure cells) right under her nose. How do they build a queen cell versus a worker bee cell? It's all in what the bees choose to feed the egg after the queen has laid it. Every egg gets royal jelly for several days after it is laid. The egg destined to be the queen (who knows why one egg is chosen over another!?) continues to be fed royal jelly all during her development. That exclusive ingredient, royal jelly, is what makes the queen cell so big. The ancients believed that the royal jelly was what made the queen live so long in comparison to the worker or drone bee.

Royal jelly is an amazing substance. It is secreted by the bees who are tending the larvae. During the time the average worker spends in the hive tending larvae, she is called a "nurse bee." Royal jelly is a milky white substance secreted from glands on top of the bee's head (hypopharyngeal glands). I've heard royal jelly is on sale at health food stores as a rejuvenative, anti-inflammatory and antibiotic. It is in moderation good for us, and is certainly loaded with Vitamins (B-1, B-2, B-6, C, E) plus niacin, folic acid, hormones, minerals, enzymes, pantothenic acid, etc. ALL BEE PRODUCTS ARE GOOD FOR US. THE HIVE HAS BEEN CALLED "THE OLDEST PHARMACY IN THE WORLD."

The eggs chosen to be workers or drones are fed pollen and honey until they emerge. Interestingly, for those 2-3 days that the worker eggs are fed royal jelly their weight multiplies 250 times! The queen is still much bigger than they are (42%) and bigger in most cases than the male drones. Plus she lives 3-6 years and the worker lives only 28-40 days. (But I do think the limited life span is because the worker bees are out in the world working every day plus they work in the hive constantly. They literally work themselves to death.)

Back to supercedure aka gyne aka queen cells. I've seen them on the bottom of the brood and they are many times the size of the little worker's cell. Some say they build them in the middle of the brood comb, but the only ones I've see have been hanging in the bottom of the brood comb.

Now if a queen is old and not laying as well as the workers think she should or if she doesn't give off enough pheromone (smell), they will just kill their monarch! Many human monarchs/rulers have suffered the same fate. Loved and then killed (e.g. Julius Caesar). Sometimes they will sting her to death. Often they will kill her by surrounding her in a ball. The crush of bees gets very hot. She dies of overheating essentially. This is called "balling the queen." I've never seen that. Maybe I will.

But they do always have a hedge against a queenless hive with these supercedure cells. I think it's prudent.

When Larry gets a new queen, he has a device like a little, enclosed clip board that he puts her in. In order for the bees to become familiar with her and her smell, he hangs her in his contraption out on the clothesline and all the bees hang from it in a great beard of bees. He says they won't sting you "for anything" when they're hanging there in their beard.

Last year Larry saw a good swarm in the top of a 70 ft. fir tree. He, only 12 years old, climbed the tree, sawed off the branch with the swarm on it and shook all the bees into a plastic garbage bag. To make sure he had the queen in there, he got all the little clusters of bees on the adjoining branches. Then he fastened the bag with a tie strip and had himself a new hive---free.

He paid $1,800.00 for his 60 hives from a man there in Uniontown, Pa. They were used hives and "were mostly dark green inside." Larry lives out in the back country near Ohiopyle Falls. He has mainly Italian bees like I have: "They're mean, but they're the best workers," he says.

Nice little interlude with Larry. A brave, industrious young man. A fellow fascinator.

May 5, 1982

Everything has been so hectic. I'll try to catch up.

THE QUEEN: She appeared on the right side and ambled in an arc across the hive; laid one egg---27 seconds; exited. That was 5/2/82 when our Kathy, age nineteen and away at college at Pitt, came home for a visit. She is the queen of our house.

The queen doesn't lay on this side anymore that I can see, but she does appear frequently, ambling slowly over the comb, stopping occasionally to inspect the stores of honey, capped and uncapped. Everyone continues with their work when she's around, but the workers do clear space for her in front of her and on all sides of her. I have observed that before. Also, 6-13 workers circle her if she happens to stop for a while. I feel these are Inspection Tours. She's checking to make sure the honey, pollen and larvae are all in order. She probably gives commands if she finds anything not to her liking. Sort of like an executive touring his plant and finding this and that awry or in need of attention. Her laying seems almost over for now. She must make sure there are correct provisions for the developing brood, her workers and herself.

HONEY: Noticed the first CAPPED honey the morning of 5/3/82---just nine (9) days after we had noticed the first honey glistening in the comb. So maybe it takes 7-9 days for the nectar in a filled cell to become ripe enough to cap? We can SMELL the honey in the kitchen. It smells like honey with a slight hint of fermentation. When I had my Medieval Feast party, Paul L. brought a bottle of mead (fermented honey) from New York City. Nobody had tasted mead. Even I, a teetotaler, tasted it. It seemed to go well with the medieval recipes I had everyone make. We sat around in the Carriage House with a fierce fire blazing just reading and talking about medieval lays.

The first cells were capped on the upper right hand side. As of today, a lot of the upper right, some in the upper middle and a lot in the upper left of the comb are capped.

The general layout of the <u>lower</u> comb, and there is a pattern here on this brood comb, is: Upper part of comb is honey. Middle of comb is pollen stores. Bottom of comb is developing brood. The demarcations are wavy---again that HILLY type of rolling pattern I noticed when they were laying comb. There is a portion of honey, larvae and a small amount of pollen all together right smack in the middle of the bottom comb.

All of the pollen on this side is various shades of golden. Really luscious looking. The blue pollen of the Scilla siberica that I observed them bring into the hive is not on this side.

Saw one of my bees inside one of my big red tulips, believe it or not. There are hundreds of them on my azaleas, the white ones and the pink ones. Azaleas are poisonous and azalea honey would, I imagine, not be good for humans. They, however, seem to feel it will be just fine for them and the uncapped brood.

Today I heard a terrible buzzing sound by a white hyacinth. I couldn't find the bee in the flower anywhere. The buzzing continued. After searching, I found her stuck between two leaves right at the bottom of the plant. She couldn't get out for some reason and was crying out in frustration hoping for help. I bent back one of the leaves and she gratefully flew away. The hazards of foraging.

Have noticed they like the purple flowers of the ubiquitous ajuga.

LARVAE: Today they were capping many of the larvae cells on the lower left and middle of the comb. They seem to like to work everything from the corners inward---laying wax, storing nectar, capping brood. For days I have seen the little workers upside down in the cells where the brood is developing with only the tips of their abdomens sticking out. They have spent hours like that in just one cell! They must be feeding and tending to the tiny egg until it reaches a certain point. Then after all those hours of attention, they know it is ready to be capped, that it is strong enough to develop the rest of the way on its own The whole process from the laying of the egg to the emerging of the bee takes 16-24 days.

When the workers were fashioning the comb, they would go into the hexagon, but not ALL the way in as they do with the developing brood. Beautiful, tiny, black bottoms against the orange-white comb! Yes, orange. Just yesterday afternoon, the comb turned from white to an orange-ish cast. Just in a matter of hours! Maybe it has something to do with the fumes of the honey at a certain point or---? The wood in the hive is also beginning to assume that orange-gold cast.

Certainly GOLD is the essential color of the honeybee. Almost all associated with her is or becomes GOLD--- the sun first of all, the honey, the comb, the wood, the pollen, half of her natural coloration. Plus the queen's abdomen (origin of all bees) is a dull gold like a good, old Spanish galleon ducat. Saw recently a picture of a gold ducat minted in the Netherlands when King Ferdinand and Queen Isabella were rulers over that area. The two monarchs were facing each other pointy nose to pointy nose. Each had on a crown that was the same size as the other's. Now that's a couple who were really equal! No drone that Ferdinand. What a queen that Isabella.

I have had my bees for 18 days now. They have made an entire civilization in that short time and have forever woven me into their lives. Thank you, Lord, for Apis mellifera! ("Apis" meaning "bee" is the genus. "Mellifera" meaning "honey" is the species. Got to remember these things, Sandy.)

May 7, 1982

Not much of the honey on the brood comb is capped. Just a small, intermittent strip across the top of the comb.

I can now see the larvae! I'm going to have Blake draw one because he can do it better than I. About three-quarters of the larvae are capped. There are some bigger cells in the lower left-hand side of the comb. I don't know enough to know if they are supercedure cells. Of course, it was these same queen cells that my friend in West Virginia, Larry, always tried to find and destroy. He missed one and they killed the queen he had bought because they already had "built" a new one. Are my bees secretly fostering this queen's replacement? What drama in such a small space.

The mature larvae are curled up in the hexagon. They are big and fat and glistening enough to be seen with the naked eye, sans magnifying glass. Small, medium and large larvae are all over the brood comb now except in the top rows. Now I can see that almost seven eighths of this side is larvae in various stages of development from teeny to already capped. They are glistening white and beautiful, delectable!

When I use the word "delectable," it has edible connotations. Some biologists have suggested that bee larvae be considered a "product" of the hive. Just harvest the undeveloped bees and use them as animal feed or for human consumption. I know that all insects are edible. They have a higher protein content than beef and are considered by those who consider such things as the greatest source of "untapped" protein on the planet. In Africa, parts of Asia and South America people regularly eat insects. In Malaysia they roast and sell on the street huge insects the size of dessert plates. Here in America we don't eat insects. Never have had to, I guess. We, also, abandoned the European practice of eating blood when our ancestors came to the "New" World. There is a restauranteur in our town from whom one can buy "beaten blood." For blood pancakes and sausages, I guess. I know all the Swedish au pairs LOVE and long for their Svedish blood pancakes.

Do you have an ethical problem with eating bee larvae, Sandra? They would have to be aborted. That's bad. They, however, are not human beings. If you could survive by eating aborted bee larvae, would you choose to eat them or to die? I'd probably choose to eat them. Another ethical question. If I went down in the Andes in a plane and survived, would I eat another human being to survive the survival? DEFINITELY NOT IF THE HUMAN BEING WAS STILL ALIVE AND WOULD HAVE TO BE KILLED TO BE EATEN. Would I eat a dead person in order to survive? If I had exhausted all the materials on the plane and the abundance of life about me, I would have no compunctions MORALLY about eating dead people in order to survive. "They" are not there. I would have physical and emotional compunctions. I've told my children that if we would be in an isolated place and without food or rescue and if I was dead, I would want them to eat my flesh in order to survive. But I know myself well enough to know that I would NOT kill another human for self-survival. I've read all the Donner party-type books and that's not me. But bee larvae for survival---definitely. Bee larvae for animal food, nutrient content, whimsy food---definitely not!

Note: In Korean and probably other Asian markets they sell cans of silkworm larvae. Some use them as fish bait, but others are in a sauce seasoned with ginger and garlic. I doubt they are feeding sautéed and sauced larvae to the fish! People obviously like these delicacies.

In this book on this page Jesse drew me with my Ben Franklin glasses on watching my bees and saying, "Oh, oh, oh, Jesse..." At least he, young as he is, has captured the awe of it all.

May 12, 1982

Several drones buzzed me as I was out in my garden today. I noticed them noisily flying from the spent poppy leaves to the leaves of the Campanula glomerata nearby and from there to the leaves of the mums. It struck me that perhaps they reconnoiter plants for the workers. I.e. they identify plants, knowing in some way when they will be ready for nectar/pollen gathering. Then they relay this information to the workers?

Under this scenario, drones would be like commandos who go ahead and get the lay of the land for the landing troops or like front men who go into a city and stage everything before the evangelist moves in. I really feel that DRONES MUST have other functions than mating and morale!!! And, yes, hives without any drones are "dispirited," as they say.

May 16, 1982

The comb is FILLED with brood in all stages of development!

Nineteen days ago, 4/27, we noticed the queen first laying on this side, so in several days the bees should begin to emerge from their waxen shells.

1:30 A.M. I'm observing late into the night. The activity in the hive never ceases. No rest for the weary it seems. A worker bee was outside the hive on the window flying aimlessly around. I shone the flashlight on her and discovered she followed the light. So I directed the beam toward the bee gate and she found her way back into the hive. Wonder what she was doing out this late? She must have been waylaid by something. Not coming home until 7-8 hours after old Sol has set! Unheard of.

The hum is very audible. I think they're excited about the bees that are going to emerge.

For the four or five hours when the sun hits this window directly, hundreds of bees play up and down the whole window. I KNOW they're luxuriating and FEEDING off of the sun and are not simply trapped in the storm window part. When the sun is not on the window, they fly in and out with perfect precision. The glass plus the sun must be a wonderful warmth.

I love to stand outside away from the bee gate and look up into the air and see them, like pieces of soot, flying home and making that right hand turn into the hive! Most who exit the hive take a left and then another left over my back porch. But several weeks ago, they took a left and then headed right over the carriage house. Must have been some important source they knew about in that direction.

They love to forage the tiny flowers of the purple ajuga growing in profusion in my grass and, regretfully, in my gardens. (I am constantly finding a thread-vine to pull them out. They are very tenacious in spite of their delicate vinework.) Anyway, my bees have given me a new appreciation of the ajuga's value to insects. They loved the blue scilla, too. And they took to the lavender azalea when it first bloomed, but then after several days left it to the hungry bumblebees.

Those furry guys really love azaleas. Wonder if they drive away the honeybees or if each has its season with the azalea? I've found my little ones on all the teeny white flowers that intermingle with the grass---flowers so small that it takes my magnifying glass to delineate their centers and petals. They like these tiny flowers. Except for pollen gathering, they virtually ignored king tulip. No Tulipomania for them. Never saw one of them on the daffodils. A few on some hyacinths. None on the Phlox divaricata. They must be getting all this nectar in other yards or from all the small wildflowers that they have made me notice for the first time. I hope they like my June bloom garden!

2:00 A.M. Just saw my first bee being born!! I was ready to go to bed about 15 minutes ago and noticed one of the cells was cracked. I watched the bee eat her way, pick her way out of the cell. It took her about 10 minutes to be "born." She emerges shaky, moving, flexing her wings. The new bee is a dull color, not the hard black and gold markings of the others. Her abdomen is almost all a dull beige. She starts moving haltingly like a newborn colt unsteady on its legs, moving UP the comb. One bee is feeding her from her proboscis. The others ignore her as she falls on her side several times stumbling up the comb. She gets to some honey and feeds---then continues unsteadily DOWN the comb. I'm writing as I watch this miracle of birth!

2:15 A.M. Saw another one emerge. Same shaky unsteadiness. I notice they both had a gold spot here. Both kept rubbing their legs against their abdomen---for grip, I think, or maybe to get off womb stickiness? They both found it hard to get a grip on the capped brood and by rubbing their legs on their abdomens were able to get some traction. This one, also, headed UP the comb after a few false starts down and sideways. That's, of course, where the honey is. Flexed her wings, also, and the same dullness of body---not hard, urgent stripes like the mature workers.

The queen is on this side---laying brood, allowing her circle of admirers to smell, groom and touch her. She's done a great job and deserves a rest that I'm sure she won't get.

Imagine. A good queen bee can lay up to 3,000 eggs a day during "laying season." She starts laying several days after she is mated. Ideally, that should be in the early spring so that the brood can forage in late spring. She is at maximum production during the summer. During the Fall, egg production declines because winter is coming.

Someone calculated that every day she lays, she produces more than two and a half times her body weight in eggs! Day after day after day....

2:20 A.M. A crack in a worker cell.

2:35 A.M. She's out. Three quarters of her came out in about 10 minutes, but one quarter remained inside the cell. A worker came by and fed her some nectar. This particular one had to be helped out of her hexagon. Some other bees came over and chewed away some of the cap. They ignored completely another bee that was nearby and trying to emerge. She had to get out all by herself without any assistance. Why? Maybe they knew this one needed help and the other one didn't?

May 20, 1982

The queen is laying again in the cells where the previous brood emerged. It takes about two days for the workers to clean up and ready the just-used cell for a new egg. She's now laying fast! Blake and I timed her: 9 seconds; 8 seconds; 7 seconds; 10 seconds. When I first began observing Laying Time, I felt I could calculate roughly how long it took the queen to lay an egg. Now I feel it's a pretty arbitrary thing----as long as she wants to linger. Maybe it's similar to delivering a child. With one birth it can take you eighteen hours; second child---six hours; third child---two hours; fourth child---ten hours....

May 23, 1982

This is Meme's birthday. My dear grandmother, how you would have loved to watch these bees, Meme. As a child, summers, in Zaleski, Ohio, I remember you in the early morning dew out among your flowers. Those hollyhocks! They were so big my eight-year old head could fit in the trumpet blooms. You'd throw the slop jar on your garden and what flowers you grew. Maybe my late blooming love of gardens and bees and things natural originated in those wet summer southern Ohio mornings.

Thousands and thousands of newly emerged ones make the combs golden with bees. Can hardly see any brood through the press of bees. Thousands are in neat, irregular rows on the bottom of the brood comb. A sign of impending swarm? What do I know?

It's been raining for four or five days and it's 47 degrees, so they have been hive-pent by the rain and temperature. The honeycombs are full and they constantly feed from them.

I feel this is an optimum condition hive:

1. Good laying queen. Has laid again in all the cleaned cells.
2. All the upper honeycomb is now filled with nectar, i.e. good nectar flow.
3. Much pollen all over the brood comb.
4. Syrup water for surplus. They're taking one bottle a day with all the new bees. They were down to half a bottle a day.

The queen obviously originally laid in the bottom of the brood combs because those were the first to hatch.

Can I draw the conclusion that the queen first lays in the bottom right and the bottom left hand corner of the brood comb? Then she proceeds inward to the bottom middle? The extreme upper right hand corner has always been for capped and uncapped honey stores to feed developing larvae. It has NOT been favored by the queen so far.

Have see the first few drone babies in the brood combs. These are the first drones I've ever seen in the brood part of the hive. The original drones from the package of bees ALWAYS stayed up in the honeycombs. They ambled around, fed and were fed, went out to fly (perhaps scout FUTURE nectar flows?), died, were removed.

On a gloomy day like today, the bees:

1. Are not making the characteristic hum. The hive is relatively silent except for an occasional flurry of communication that spurts, static-like, from the vent holes on the sides of the hive. Is the HUM then a sign that they are content because nectar, pollen, propolis can be gathered? Do they work happier, i.e. hum, knowing others are foraging and thus perpetuating the hive. And is the possibility of, the feel of, the presence of Old Sol an additive to the HUM?
2. They huddle closer together, possibly for warmth and comfort, possibly because the hive is so crowded---everyone's packed into the city.
3. Normal activity doesn't cease at all, but the spirit that the sun and warmth elicit is lacking.

Note: Have to finish up on the Beehive Cluster, Utah's astronomical symbol. Its correct name is Praesepe, Latin for "manger." I couldn't find out why they call it the Beehive Cluster. Most stars have an "M" number.

The Beehive Cluster is M44. I looked up the M part. Lots of star clusters are called M22 or M7 or M some number. The M is in homage to Charles Messier (1730-1817), a French astronomer. Young Messier became fascinated with the heavens at the age of 14 when he observed a six-tailed comet. In 1748 he saw his first solar eclipse. He devoted the rest of his long life to searching the night sky for comets. Along the way, he identified 110 fuzzy shapes and catalogued them. M1, his first entry, is the Crab Nebulae. These nebulae had been known to mankind for thousands of years, but Messier was the first to compile a LIST of them and their locations. I'd never heard of this man. What a way to live your life. You live for the night. You wail when it's clouded over. You rejoice when it's clear as a bell. Searching, searching, seeking, seeking. "Is that something? No. What's that? Is that fuzzy thing a comet or a star or what?" Amazing man! He was the 10th of the 12 Messier children. Today in France they would have aborted him for sure. And in Germany they would have aborted Johann Sebastian Bach. He had five older brothers and two older sisters. An eighth child! Unthinkable that anything good could come of him!

May 25, 1982

Still c. 50 degrees, but when the rain ceased, the bees started to hum and 20-30 have ventured out onto the window. The general hum has resumed, too. Maybe they know it will be nice tomorrow, that this is the end of the period of rain.

The queen has been resting on the middle left of the comb for about 30 minutes, 20 minutes of which she has stayed in the same place and position. A relatively constant 12 bees encircle her. These are, of course, called the Queen's Attendants. The three attendants near the right side of her abdomen had their long, thin, red tongues out constantly stroking the side of her golden abdomen. They then used their front legs to stroke their own tongues. Are they making their tongues more receptive to the abdomen or are they putting on the legs something from/for her abdomen?

The four attendant bees at her head were constantly intertwining their antennae with her antennae. Were they conveying information about the state of the hive; the number of eggs in larval, pupal, hatched stage; honey and stores knowledge; which cells are ready to receive eggs; or was that just an old-fashioned pep talk to the quarterback?

Yes, I do believe they collaborate with her and guide her where to lay even though she personally inspects each cell to be sure. Her entourage is part of the egg-laying team and is vital to her function as Layer. The quarterback can't play the game all by himself even though he is pivotal. So the queen and her team. Yes.

She did a lot of laying in newly-cleaned, recently-hatched cells and I can see larvae in all stages of development on the comb.

May 26, 1982

Sunny and 72 degrees. After all the gloom and rain (which I personally love), they are ecstatic! The bees were right yesterday. It is a sunny day. Thousands are flying in and out; are on the window busily sunning themselves; are picking up the dead from the bottom of the outside window and flying away with them to drop them far from their hive. They are neat and very concerned about the dead polluting their living area as are we humans.

On the inside of the hive more than half the bees are dancing excitedly. So many have left the hive that I can see parts of the comb. It's about half capped with honey; half bees in the developmental stage.

Definitely in this hive she lays beginning at the bottom and going up the comb with only random laying on the top.

The honeycomb above is constantly chock full of golden nectar. The hexagons are never capped because they consume the nectar so quickly.

There must be a queen cell or two in all this mass. They'll have to swarm soon---just too crowded. The cells at the bottom which I at first thought were gyne (queen) cells were/are probably drone cells because they never achieved the characteristic peanut shape of the supercedure (queen) cell. Those larger, bulging cells that are probably drone cells are almost always built at the bottom and extreme sides of the hive. This way they don't impede traffic.

Have, also, observed them fanning their wings over a cell---a blur of wings. To cool? To heat? To evaporate moisture? I know they fan to evaporate water out of the nectar before it is capped.

The queen is laying again. Like yesterday's timing, averages c. 12 seconds per egg. When she leaves through the opening in the upper left-hand corner, two bees invariably leave with her. Are they ladies in waiting, constant courtiers?

These bees have to swarm soon. It's so crowded.

May 29, 1982

Steve and I just got back from a getaway night at the Hilton. We took a nap yesterday afternoon and I dreamed my bees swarmed and went into the garage and carriage house.

We got home about 20 minutes ago. Steve and Blake (age 16) went golfing at Silver Spring. Jesse (age 6) and I were out on the back porch talking about the therapeutic value of art. Jesse's one gerbil, Jimmy, was killed three nights ago by our cat Marty and he wrote a poem about it, about his love for Jimmy and his grief. Kathy (age 18) helped him and he now feels better. Art does help us to get out our feelings in a way that transforms the grief, pain into a kind of pleasure. Someone dies. Grief. A poem is born. Pleasure. In some of my poems and especially in certain passages in my novel <u>Snowtime</u> where so much of Meme is being remembered, I weep with bittersweet feelings. There is a profound and deep joy and satisfaction when something GOOD is created by you. Of course, there are different internal standards for what is good depending upon the literary, artistic background and talent of the writer, painter, musician. But even a created mediocrity can have therapeutic value to the creator. I am so deeply thankful, Lord, for human creativity and for the creativity You have placed within me! How often throughout my life, the writings, art and (to a lesser extent) music of the ages has inspired, lifted and solaced me. In that way we do reflect You, the Original Creator. You are our Father--- even if some of us do not recognize You as such.

Jesse said suddenly during our talk on art, "Mommy, look at all the bees!"

I looked and thousands of them were swirling in circles over and under the evergreen right here next to the back porch! We ran around to the hive and thousands more were on the window. "They're going to swarm!" I yelled. Jesse and I rushed into the kitchen to open the observation window. We could see almost all the brood comb. The bees that were still in the hive were dancing with excitement.

They're SWARMING!!

It's now 12:45 in the afternoon. They have been swarming for 20 minutes and have already formed a big cylinder of bees on a strong branch of the evergreen tree. The queen is in their midst I am sure.

Kathy is here and has made an interesting observation. (I'll have to investigate and see if anyone else has observed this.) That when they are filing out of the bee gate in the window, they are coming out in straight formation lines starting at the right and arcing to the left.

I see the activity inside the hive is proceeding normally now. There are bees cleaning cells. There are bees coming in laden with pollen in their baskets. There are bees feeding on the honey stores. The ones who have been "designated" to stay have resumed their tasks.

12:57 P.M. The number of bees on the comb in front here is increasing and all seems like normal hive activity. That means to me that the swarming is over---fait accompli. Jesse just came in and told me the swarm on the tree is gone already!

12:59 P.M. The entire swarming process took 34 minutes.

I'm outside now looking at the entrance to the hive. Thousands are flying lazily, not frantically, around the entrance and thousands are on the sill and hanging around the bee gate.

1:50 P.M. Drama and Trauma!

When I went outside and saw all those bees, it was a bad sign. Plus the inside comb was filling up pretty fast. An ABORTED SWARM.

The clipped queen! Probably. I looked on the grass outside the entrance to the hive and saw about twenty-five (25) workers wandering around in the stones and grass. I examined the little colony. There was the queen! There was the lady I had picked up and put back into her queen cage on hiving day. She, of course, was shipped with her wings clipped so that she couldn't lead out a swarm. You don't want your bees to leave if you have outside hives because then half your workers go and you only get half the honey. However, I'm just observing them naturally. But she didn't come naturally. She came with clipped wings. I'd forgotten about all that, but I've got SO MUCH TO LEARN.

What happened was: she led them all out because it was too crowded. It is nature's way of preserving a hive and starting new ones. It's a propagation technique. But she couldn't fly which I'm sure she found out as soon as she tried to spread her wings to lead the swarm out of the hive! Down to the ground she went. So when they had all gathered there in a giant beard on the evergreen branch, they realized they couldn't SMELL her in their midst. Where did she go? Where is our Monarch? She must be at our hive. Let's go back and see.

A swarm won't leave without a queen, so they all flew back into the hive again. Some believe that when they leave the hive in a swarm, they immediately "forget" the entrance. This, of course, is not true otherwise they who hung on the tree for about 20 minutes couldn't have found their way back into the hive again. And they all did except those faithful handmaidens who were wandering around on the grass and stones under the bee gate with their ill-fated monarch. And she is a dead ruler. For the Queen who cannot swarm with them is useless. They will sting her to death and build a new Queen.

I called crotchety, cranky, lovable Ed. "My clipped queen is on the ground outside the hive, Ed. She tried to lead them out and obviously couldn't. They hung for a while and then all returned to the hive. What should I do?"

"I can't talk about it now! This is a holiday!"
"But, Ed. What do I do?" (What holiday was this?) Bored and mad Ed began. "Pick her up and put her back in the hive. They'll build another queen in five days and swarm with her."
"Pick her up and put her back in the hive with 50,000 bees guarding the entrance! How do I do that, Ed?!"
Ed was becoming warmer, more patient with me. "Get a cardboard box. Put holes in it. Pick up the queen the way you did before and get some of her attendants with her. Put them all in the box. Hold the box up to the bee gate. She'll go in."
"I can't reach the entrance to the bee gate. It's 10 feet off the ground." I was really hoping that Ed would offer to come and do this for me. It would have been rude to offer him money, but I would have paid a lot.
"How do you reach the bee gate?!" he yelled. "Get a ladder!"
"Ed, this may be hard to believe, but I don't have a ladder." I protested.

"Look, Sandy, this is my day off. I have work to do!"
"Okay, thank you, Ed."

I turned to MY trusty attendants, Kathy and Jesse. "Get a cardboard box and punch holes in it."

They got a shoebox and with a pencil punched holes in it.

When Steve and I had hived the first time, I had borrowed two bee veils from Ed. But now I had nothing to put over my head to protect me from those 50,000 bees who might not know that I was at the entrance of their hive just to give them back their queen. They ferociously guard and defend the entrance against all intruders---human, animal or insect.

I have a see-through, lime-green nightgown. Kathy tied it over my head with a piece of rope. Slim protection, but I felt a little more secure. I got a very high stool and positioned it against the house under the window. Then I picked up the queen with the mark designating her as queen on her back and put her in the box. I remembered he had said to get some of her attendants. Forget attendantS. I picked up one buzzing attendANT and threw her in the box.

I was really shaking and praying. Hold the box in right hand. Hoist self to a standing position on the wobbly stool. When I straightened up, I was face to faces with what seemed like millions of bees! Most had not gone back into the hive after the aborted swarm. They were hanging near, around, over and under the sill in a dense, moving mass. Don't think, Sandra. Just act. I emptied the box carefully on the inside of the sill. She and her lone companion tumbled out. I kept very cool. I was NOT going to have to repeat this exercise! The bee mass began to stir. I still wasn't going to jump down until I was sure she was going into the hive.

They were buzzing all around me examining the lime nightgown and my hands, but they were not stinging me. As I watched, I kept my eyes on the white dot on the back of the queen. She moved slowly around the sill, up to the gate and disappeared into the hive.

I jumped down. My legs buckled. Thank you, Lord.

So. They're all back in.

It's interesting that I had dreamed the day before that they all swarmed into the carriage house. They did swarm and they did, indeed, all go into a house again. Their own. Often dreams are prescient. I tell people that their dreams are their friends. Especially the bad and scary ones. They are trying to warn us of things in us or around us that keep us from developing into sane and effective human beings. But dreams are prescient, also, as was my dream of swarming. In the Bible dreams played an important part in the birth of Jesus. Joseph was told in a dream to take Mary as his wife even though she was pregnant (Matthew 2:20) After Jesus' birth when Herod was going to kill Him, God told Joseph in a dream to take Mary and Jesus and go to Egypt (Mt. 2:13). A dream told Joseph to leave Egypt and go back to Palestine (Mt. 1:19). When the three of them returned to Israel, God told Joseph in a dream not to settle in Jerusalem, but to go to Nazareth (Mt. 2:22). Even the Wise Men who had come to worship Jesus from Arabia were warned IN A DREAM to bypass Herod and go home another way (Mt. 1:12). In the Old Testament another Joseph had dreams that were prescient, but they got him into a lot of trouble. He dreamed when he was 17 that he and his 11 brothers were binding sheaves in the field and "my sheaf rose and stood upright, while your sheaves gathered around mine and bowed down to it." Joseph, of the many-colored coat, was already the favorite child of his father Jacob. This dream pushed the brothers over the edge. Joseph was not just a brat. He was a megalomaniac, to boot. A caravan bound west was passing by and they sold the "daddy's boy" into slavery in Egypt. (Genesis 37) But the dreamer Joseph, also, had the gift of interpreting dreams. He rose to be Secretary of State under Pharaoh through interpreting Pharaoh's troubling dreams one of which warned of a famine in seven years. His final reconciliation with his brothers is one of my favorite stories in the Bible. There was, as Joseph had interpreted Pharaoh's dream, a terrible famine. Joseph's brothers came to Egypt to buy grain. They are taken to Joseph who is 2nd in command in all of Egypt. He knows who they are, but they have no idea that this powerful Egyptian is their brother. He toys with them. Finally Joseph "could no longer control himself before all his attendants, and he cried out, 'Have everyone leave my presence!' So there was no one with Joseph when he made himself known to his brothers. And he wept so loudly that the Egyptians heard him and Pharaoh's household heard about it. Joseph said to his brothers, 'I am Joseph! Is my father still living?' But his brothers were not able to answer him because they were terrified at his presence." (Genesis 45)

I love the stories of reconciliation in the Bible. The Prodigal Son is the most famous one. I LOVE the story in Genesis 31 and 32 when Esau is so gracious to tricky Jacob who cheated brother Esau out of his birthright. Most of the broken relationship stories I hear and know are BETWEEN FAMILY MEMBERS! Just today a woman told me that her husband has a brother who cheated them out of money in business. Joseph and his brothers. The Prodigal Son and his father and his brother. The twins Jacob and Esau. Hate and enmity in families goes back to Cain killing his brother Abel!

Swarm observations:

1. Normal activity within the hive goes on concurrently with swarming excitement. An under layer of bees did all the regular hive work and seemed totally unfazed by the flurry around them.
2. How do they designate who goes out and is caught up in that glorious excitement and who stays in the hive and totally ignores it?
3. Saw more bees with filled pollen baskets on the brood comb than I had ever seen at one time. Maybe they knew the responsibility was on fewer shoulders.
4. In the fury and press of swarming, many bees (c. 20) broke into our kitchen. They found a weakness in the stripping which Steve had put around the sash of the window and lifted it up. They were bent on getting out of the hive any way they could. I opened another window and they found their way outside again. Others came in and found their way out through the open window. First time I've had a honeybee from the hive in my kitchen.
5. I saw NO hostility to me, Jesse or Kathy during the swarming or even during the time I was re-introducing the queen into the hive. Our cat Marty and dog Scrubby were very frightened and insisted on being inside the house during the swarming. Even when I was picking up the queen on the grass, her attendants made no threatening moves toward me. When I put her on the sill, the thousands didn't threaten me. Their attention to me seemed mere curiosity. Truly "swarming bees are SWEET BEES" as the saying goes!

6:05 P.M. Day of Aborted Swarm. Hive activity is normal. Queen is on this side roaming around. As a matter of fact, she had been predominantly on this side for the past four days. Not laying, just roaming. Is that inactivity typical of a queen before a swarm,

I wonder. Or was she checking to make sure she left the hive well-stocked with honey, pollen, larvae? Or was she spreading the word of the swarm? Or was she resting up? Or??? I never saw her laying eggs any of those preceding days. She was rarely attended by her circular retinue. She roamed free with a random bee smelling her, following her.

Ants, black ones, are roaming around the outside of the hive here seeking entrance to the honey and---certain death.

Truly the excitement of a Swarm is like Derby Day.

If they do swarm with a new queen in five days, that'll be Friday, June 4.

Here are some calculations I've made as to the number of possible bees in the hive at this time.

I estimate, in sum, there are c. 40,000-50,000 bees in the hive right now. I could be off by 5,000-10,000 because there could be that many on the bottom of the hive and on the sides of the hive. Plus they are often three-bees-deep in the hive. I've calculated using the hexagons of brood and honeycomb and by visual observation.

Let's see: started with c.10,000 bees. In 43 days have 40-50,000 and the brood combs will be ready in a week or two to hatch their second batch of brood. Very prolific. If you started with a town of 10,000, and in a month and a half the town grew to 50,000. AND it was kept neat, tidy and efficient. That would have been impossible for humans. In our world that would be considered an invasion or a mass exodus from one place to another with all the chaos of feeding, housing, plumbing, crowd control and disease attendant with rapid growth of that magnitude.

A Cosmos in my Kitchen

During the Gold Rush in California in 1849, San Francisco went from 1,000 people in 1848 to 25,000 people by the end of 1849. The pictures all show a dirty, mud-filled, rutty town with gun-toting men and rowdy women. We humans are just not equipped like the honeybee is to deal with invasions en masse.

When the bees go to get a grip on the glass, I can see their whole underbelly. The bottom legs are shaped so that they grip well. The middle legs don't have the pollen baskets, so they have trouble gripping. Thus, the bees pass the middle legs through their back legs to get friction, traction, a better grip.

They, also, use the fur on their thoraxes to "clean" the legs to get better traction. They especially rub their front legs on the thorax fur.

I sure wish I knew what all their hive dancing is about. I've seen them dance with laden pollen baskets. Is that the "I've-got-pollen-dance?" Pollen is one of the sine qua nons in the honeybee hive. It's the primary food of the developing brood. Its nickname is "Bee Bread."

Now at midnight some are doing a Shaking Back And Forth Vigorously Dance for no reason I can fathom----yet.

They look so sweet, baby-like, vulnerable all lined up at the top of the brood comb in a row---silent, dull-colored, resting with their little proboscises and front legs on the wood between the honey and brood combs. Some bees are totally at rest. Some flex their antennae from time to time. Some move over and let another one come in.

They rub their legs together for friction as we do our hands.

June 1, 1982

Every day since they attempted the swarm on Sunday, it has been rainy and cold. It's so congested in the hive that even in the rain, many are outside on the bee gate.

The drones on the brood comb have a much harder time getting a grip on my observation glass than do the workers. They flail around and finally slide to the bottom of the honeycomb. The workers are persistent and rub their legs together, try traction, rub legs on fur to try to stay or to climb on the glass. But the drones will try, fail and fail again without trying any innovative or observed methods of getting traction.

There definitely is a "POLLEN DANCE." I've observed it with almost every worker who carries pollen on this side of the glass. It goes like this: she shakes her whole body vigorously from side to side; walks away; shakes again and often goes in a circle shaking side to side. THIS PARTICULAR IN-HIVE DANCE IS ONLY DONE BY BEES WITH FULL POLLEN BASKETS. The pollen itself is usually deposited on combs I cannot see. But her full-pollen-baskets-dance is real.

Other possible explanations for this dance:

1. She's announcing replenished stores.
2. She's just plain happy.
3. She's calling attention to the pollen as one does to a trophy.
4. She's laden and that makes her want to dance or need to wiggle in that way in order to move, mobilize, etc.
5. It's another one of the directional dances. "Here's where I got the pollen". But no one around her is

A Cosmos in my Kitchen

leaving to get pollen. No one seems to be heeding her directions.
6. None of the above.

On a lone part of the brood comb I see a possible queen cell.

June 2, 1982

They haven't killed the clipped queen. I've just seen her. It's 11:40 A.M. I sure do hope she doesn't try to lead them out again! Leave her here and alive and go out with a new queen. That's an order from your bee mistress Sandy. I'd hate to have to face you all again!

She's not laying, just roaming. Bad sign. Her abdomen is much thinner.

They're bringing in oodles of pollen and doing the I've Got Pollen Dance.

More than half, maybe three quarters of the brood comb is capped with the second batch of brood. She is prolific. Again the top part (about 8 cells down) is filled with pollen, nectar, few larvae. Same pattern as last time.

3:50 afternoon. Beautiful, sunny, warm. They're swarming again!! And probably with the clipped queen again. That explains why I saw so many more pollen carriers on the brood comb than ever before and, also, the nectar combs were filled with bees engorging themselves (as they do before swarming).

Look at those pollen carriers. Thousands of bees are exiting the hive and are in full swarm excitement and chaos. But the bees with their pollen are fighting to get INTO the hive as though nothing extraordinary was going on. They battle the exiting bees for space. It's like they are going Up The Down Staircase.

Some few bees on the nectar combs seem to be feeding voluptuously for the journey. They all leave with their stomachs loaded with honey. The honey is all gone from the whole right hand side of the honeycomb.

The glass is very warm to the touch with the heat of swarming. Wonder how high the temperature gets in the hive during swarm activity? Seems to be about 8-10 degrees higher than usual. Usual in-hive temperature is about 91-96 degrees F.

More than half the hive is gone and I can see the brood comb. How beautiful. Burnished, golden brown with a high patina.

They've all congregated on the same evergreen, but on a different branch. It's now 4:07. We're 17 minutes into the swarm.

4:11 P.M. Three quarters of the bees in the hive left with the swarm. Some are still bringing in pollen. Some are cleaning cells. The rest are running crazily all over the brood and honeycomb. Calm and craziness reign in the hive. They all know what's going on, but the pollen carriers and the cleaners/feeders are still diligently at work as though nothing were happening.

On The Swarm Site: The bulk of the bees are in a beard on the heavy, yew branch. When they congregate en masse in a lazy and droopy OR tight and compact formation, it's probably called a "beard" because it does resemble the types of facial hair men have sported over the centuries. Wild and wooly, High and tight. Might be wrong about this. But many of the bees are flying in seemingly random circles, ellipses really, and in zigzags all around the top of and underneath the evergreen.

4:15 The brood comb is filling up again. I can't believe it. They swarmed again with the clipped queen!! And I'll have to find her and...

4:45 Put on my jeans and tucked them into boots. Put on a turtleneck. Covered my head with my light green see-through peignoir. Tied it around my neck with a bathrobe belt. Found her floundering in the grass. Picked her up---no attendant this time--- and put her in the shoe box. Got up on the stool. I had left it there just in case this might happen again! Faced those thousands of buzzing bees---again. I was three inches from them. Fighting to see through the lime veil of the nightgown with the bees swarming all over it, I dumped her highness on the windowsill. Would she trudge her white-spotted body up to the bee gate and enter? Yes, she entered. Got down. Wasn't stung. Maybe they know I'm their keeper?

They do operate on pheromones, smells. Many believe they know their keeper's smell and won't readily sting him/her. Maybe that's the explanation. If they had smelled my pheromones those two times, they would definitely have gotten a proboscis full of FEAR!

Entire re-hiving of the queen process---30 minutes including donning of my exotic gear.

6:00 before dinner. The first bee sting since I got the hive. Jesse was in the kitchen getting a glass for iced tea. One of the bees that got in during the swarming flew to his neck. He didn't know what it was and brushed his neck. It stung him. He came out crying, "I think a bee stung me, Mommy." I comforted him. Put a poultice of baking soda and water on the sting in order to draw out the venom. He recovered within minutes. He's a brave little one. We went in the kitchen. The poor bee was lying on the floor. I brought her outside and Jesse and I watched her drag around on the cement with a white, sticky thing protruding from her abdomen. The stinger and part of her abdomen had come off in Jesse's neck, so this was probably intestinal residue. She dragged on the concrete until the white came off and then was airborne, tentatively and low, c. 4 feet off the ground. She flew away to her imminent death. The honeybee will only sting if she fears her life or the life of her hive is in danger. She only gets to use her stinger once in her lifetime because it and part of her abdomen come off when she stings. When Jesse casually brushed his neck, she felt he was after her and stung him. If you accidentally step on a honeybee in the grass, she will sting you for the same reason. But honeybees are not aggressive like some other bees and will only sting if they or their hive are in danger, either real or perceived. Now the big male drone for all his loud buzzing and humming doesn't even have a stinger. The queen, on the other hand, has a big stinger and will use it, but her stinger doesn't come off and she can sting over and over again. People who have been stung by honeybees should realize that for every sting they receive, a honeybee dies. Anyway, as Meme used to say, "Bee stings are good for you." She may have been right because lots of people with rheumatoid arthritis, tendonitis and multiple sclerosis are using bee venom for relief of symptoms. Some of them even raise their own bees so that they can bring a few into the house a couple times a week and goad them into stinging their knee (or whatever) joint. I'm sure the western doctors won't be interested in this kind of alternative therapy. But there are interesting testimonials about the relief some have received from this unconventional approach.

6:50 P.M. Several bees are hanging around the area where the bee was mortally wounded by stinging Jesse. They smell that one of them was in trouble in that 3 ft. square area. Often if one bee stings someone, it is her "death" smell that draws other bees to inflict stings, also.

They're all in the hive again and acting congestedly normal. I hope they don't swarm each day with the clipped queen. They have to know by now that she can't lead them out! Besides I've had enough derring-do for a while. Love that phrase "derring-do" which comes from the Middle Ages and means "daring to do." Edmund Spenser, he of "The Faerie Queene," made it synonymous with "masculinity."

June 3, 1982

11:20 A.M. Some Conditions for Swarming in a Normal Hive:

1. Congestion.
2. Hence, it is too hot inside.
3. Warm, sunny outside. Hence swarming occurs in spring, summer, early fall.
4. Enough nectar to engorge themselves and still leave enough for those remaining.
5. Enough pollen so they are physically healthy and enough to leave as stores for the ones who remain.
6. A viable queen to lead them out. A bought, clipped queen will result in an aborted swarm, as amply demonstrated.
7. Sun must be "out." The first attempt on Sunday was during a one and a half hour patch of sun in an otherwise dreary day.

8. Temperature at or above 65 degrees F.
9. Brood comb left in all stages of development.
10. Enough bees to make a swarm and still leave enough to parent the hive.

June 6, 1982

2:30 A.M. Have not seen the queen since I put her back in the hive four days ago. They may have killed her or a new queen has killed her or...Haven't seen a sign of any kind of a queen.

It's been pouring for two days and they've been hive-pent. With hundreds of bir0hs each day, they are literally packed in like sardines. Every movement is sharply curtailed and modified by the congestion. No busy trips from one side of the comb to the other. No unnecessary shaking and maneuvering. In fact, when I take the wood panel off to observe them through the glass, they're almost lined up in semi-straight horizontal rows. After the cover is off for a while, they become haphazardly arranged again. Perhaps without the light in the kitchen, they do queue up in the hive?

Re: Swarming: Bees will swarm even if the day has been rainy or will be rainy. Before my first swarm, it had been rainy and bleak. Then the sun came out for about 1½ hours. They swarmed. So if the congestion is great, they'll take any sunny spot and swarm anyway. If that swarm had had a viable queen, they would, no doubt, have been hanging on that branch in the pouring rain that followed later that afternoon.

Saw one bee on this side of the brood comb fanning. She was facing the glass, fanning, so I think it was for cooling down the hive with all that congestion. She's in the center of the hive and has been the "fanner" for about 20 minutes now. I don't see any others fanning and would see them because the other bees give slightly more space to the fanner so her wings can whirl.

Note inserted later: In 1325 B.C. a beautiful gold fan with ostrich feathers was placed in the tomb of the Egyptian boy-king Tutankamen. Ancient royalty had elaborate fans and many slaves to cool them in the still desert heat just like our queen has many workers to fan her hot and humid hive. In <u>Gone With The Wind</u> there is a scene where inane wealthy plantation girls interrupt a day of festivities for afternoon naps. Young black slave girls fan them as wily Scarlett O'Hara sneaks away to declare her love to insipid Ashley. Plain, ordinary people have always fanned themselves to ward off heat and insects with fans made of palm leaves, thin wood or woven reeds. "Fan" comes from the Latin word "vannus" meaning "wind." Most of us identify personal fans with the Chinese. When I was young, our family went to a Chinese restaurant frequently. The owner always presented my sister and me with a paper folding fan. When I splayed it, colorful flowers and leaves appeared on a Chinese red background. At amusement parks in Ohio if you broke only 1 balloon with the 6 darts, you got a Chinese paper folding fan. But it wasn't the Chinese who invented those folding fans. It was the Japanese in the 700's A.D. In the 800's the Chinese embellished the folding fan by creating slats of carved ivory, mother of pearl and tortoise shell. The returning Crusaders introduced fans to Europe 400 years later.

Dr. Schuyler Wheeler is credited with inventing the first electrical fan for personal use (1882). He paved the way for ceiling fans, industrial fans, rotating fans, fans on wheels, etc. For thousands of years we have fanned ourselves, have been fanned and have fanned our surroundings. Then came AIR- CONDITIONING in the 1950's. It is a great invention, but air-conditioning is not as beautiful, artful or portable as the hand-held fan. Plus, you can't carry it around in your purse or whip it open to use as a seductive technique.

Note inserted much later: In 2003 our family went to Rome in August. It was 110 degrees day after day after day. The August 2003 heat wave in Europe broke all records and killed 50,000 people in one month. 20,000 died in Italy. Before we went, I bought a little battery fan at Sharper Image because I thought Rome in August would be about 86 degrees. Steve, our grandson Tyler and I walked around that "City That Takes My Breath Away" for a week before the rest of the family arrived. The fan was attached to the back and sides of my neck by a rather prominent, wide, black contraption called the "cooling system." It was out-standing enough to attract attention. People would snap their heads as they passed us or stare with a slight smile on their faces. I was obtuse and felt if they only knew what this was, they would be amazed and desire to have one for themselves. So I would say, innocently and loudly, "Personal cooling system." I would point to this thing on my neck and reiterate in a very pronounced way, "Personal cooling system." (They spoke Italian, of course, and no matter how loudly or distinctly I pronounced the English words, they didn't understand.) They did understand that I was one crazy American. Tyler got hysterical over this (as well he should have). It provided him and the rest of the family with great hilarity the whole month of August. Actually, just recently he recalled the "personal cooling system" story. He and I had a good laugh. (Now our children are all grown up and we have 5 grandchildren.) Back to the observations written years ago.

Saw the first fleck of wax on the glass up on the upper honeycomb. There are usually c. 8 pieces on the lower brood comb glass. They probably excrete the wax and it gets stuck accidentally on the glass. They wait days or weeks often to recover the stuck wax for further use. This is the first piece, however, on the upper comb glass. Does that mean they are going to cap some of the honey up there? None has been capped so far. They use it up as fast as it can be stored. Or are they building new cells, deepening old ones. Or the bee who put it there had been eating so much honey, necessary for the wax plates they secrete, that it just came out of her wax gland accidentally onto the glass?

June 6, 1982

11:30 at night. I'm almost positive they have killed the clipped queen. Or else she is so hemmed in by hive congestion and the three days of constant rain that she can't move.

Most of the nectar on the honeycombs is gone. They've been constantly feeding and they've consumed more sugar water than usual---three quarters of a bottle in the last 24 hours vs. one third to one half per 24 hrs.

As a human being, I am impressed by how patient they are in spite of their extreme congestion. No comb can be seen and they are pressed in two, three bee deep.

There is a constant, but stationary movement of their legs, heads, abdomens. There is little lateral or horizontal movement. They are united as one in their calm and their positions. I have a feeling they say, "I'm going to make NO waves. When the sun shines or a non-rainy day comes, most of us are going to get the H out of here."

How different are we humans. In families or cities where there is congestion or extreme close quarters, there are temper tantrums, violent outbursts, crimes against each other, insanity and mayhem.

But here in the instinctual versus the willful world, here is peace, patience, humility, unity in the face of adversity. The fruits of the Holy Spirit are at work in this instance. Not in all areas of the bee's life, but here at least. If we hadn't wandered away so long ago, if we could wind our way back this very day, how much our daily life would resemble this particular moment in the hive of the honeybee.

And just think---all essential services are still being performed in this juggernaut of bees. Tirelessly and conscientiously performed. The natural world is infinitely more PATIENT than the world of man. Trees are patient. Plants are patient. They wait FOR Time and wait ON Time. They ABIDE like these bees are abiding until Time is ripe. Patience is one of the most valuable fruits on the Spirit's Tree. Only Man tears at the blossom.

The man who brought me to the Lord, Francis Schaeffer, used to say that Nature and the Universe obey God through natural law. Animals and other living things obey God through instinct. Only Man obeys God through will.

June 7, 1982

Still hive-pent. Five days of patience and congestion. It's 55 degrees, cloudy, drizzly, damp. In fact, the rains here in Ct. in the last four days have taken NINE lives

A Cosmos in my Kitchen

and there are millions of dollars of property damages. So we're not talking about a little rainy spell. Before the rains, we had painted the pool and are filling it now with the hose. So the rains have helped us there.

Wait. I just heard the MOST UNUSUAL sound coming from the hive! It's a medium---not high or low, but more high than low---sound. It goes, "Yeee, Yeee, Yeee, Yeee." I counted 29 Yeee.

One minute after the previous sounds, another series of Yeee---18 times. I'm sure only ONE bee is making that sound.

The bees seem somewhat stirred by that sound. The noise is definitely coming from within the hive and is loud enough for me to hear it without putting my ear to the hive. It can be clearly heard right in the kitchen. I did notice one bee fly out of the hive before that sound. Another or the same bee flew back in several minutes later.

I "think" the Yeee came in a series of three. The first series I heard subliminally. The second and third series of beeps I timed and counted.

It's possible the bee could make a sound like that even though I've never heard that sound before. It sounds like that "call" is forced out of the thorax---something like forcing short, rhythmic sounds from our own thoracic cavity. A combination of wind and pressure.

Haven't heard it in 8 minutes now. What could it have been?

Haven't seen the queen since I put her back in five days ago. She's dead.

5:45 P.M. Maybe this has no meaning, but the sun just peeked through the gloom---not a full sun, but sun. Were the beeps meaning, "Our scout sees the sun in a little while?" The Yeee occurred 20 minutes ago.

Note inserted under this entry one year later in June of 1983: That sound was the queens PIPING to each other and there was probably a fight. The old queen was, no doubt, killed by the new virgin queen if she had not already been killed by the workers. So the "Yeee" sound that I heard is the sound only queen bees make. The piping sound is described as "quacking" or "tooting." To me it is definitely a TOOT! There's no quacking coming from my hive.

The queens pipe to each other! What drama. An old queen hears an unhatched new virgin queen piping in her supercedure cell before she emerges: "I'm here. I'm young and I'm coming to get you." It's the age-old battle between age and youth. The present queen of the hive pipes back: "I'm more experienced than you and I'm looking for you. I'm going to find you and I'm going to tear open your cell and sting you to death." The old queen stalks the hive looking for the queen cells. She's furious. The worker bees often use their bodies to hide the new queen cells from the old queen. She tears around the hive determined to find that new queen. It should be remembered that the old queen is the mother of the new virgin queen. So in the death fight we have mother killing daughter or daughter killing mother.

A new queen or queens will emerge from their cells and pipe to each other. During the fight, the fittest wins. There is only one queen in the hive. The other or others are killed. This piping is an Amazonian war cry.

June 10, 1982

10:35 in the morning. The bees are swarming again! They have been for ten minutes. They are back on the same evergreen but on a different branch from the first or second time. Same elliptical circles swirling in the air above the beard on the branch and the same zigzag patterns crisscrossing the air below the tree. Same fury of activity within the hive. Hope this is a real swarm.

The swarm huddles vertically on the branch. The mass (beard) is about 2 feet long extending from a main 3" to 4" thick branch to an ancillary branch about 1" thick.

11:47 in the morning. The swarm is over. The swarming itself lasted c. 22 minutes, the same as the other two aborted swarms.

I have taken pictures of this swarm. I climbed up and got within two feet of them. They didn't bother me at all. I wished them a good voyage and life and thanked them for being such good workers and such good company for me.

About 50 bees fly at all times around the swarm and above the swarm. It is now 12:05. They've been hanging there c. 20 minutes.

Even though they have a viable queen (they haven't returned to hive), half as many left with this new queen as left with the old clipped queen. I can't give a true estimate, but if 40,000 left with the old queen (the hive was really empty), there are probably about 25,000 who left with this new queen.

Maybe they had more time to develop a loyalty, familiarity with their old queen. She had, also, proven herself to them by such substantial brood production over a two-month period. I really feel these are some of the reasons why the swarm is considerably smaller than the previous two. Plus the old queen was MOTHER to many of them. They were her first generation offspring. This new queen they know NOTHING about other than:

1. She comes from their old queen.
2. She is a sibling to most of them.

The old queen was a stranger to the original package of 10,000 bees. They accepted her. But the hive has grown to 50,000 and many of the original workers from the package are dead. So the old queen is truly mother to most of the hive. They share her genetic make-up.

The parent-child relationship is evident even in the animal world, so why not in the insect world? This deserves study. A start is that more exited with their mother than with their sister.

The hive is still congested so I imagine I'll have another one or two swarms this summer.

12:45 in the afternoon. Re: The bees that fly around the swarm vs. those who are packed into the swarm. I noticed there are a floating number who fly around a while, light on the outside of the swarm and immediately begin to burrow deeply into the swarm. They don't join the ones arrayed on the outside of the swarm. New ones fly off, circle and then alight and burrow in. These could be alternating guard shifts, guarding, watching from outside for danger. These guard bees would be different from the scout bees I've read about who go out from the swarm and spy out a new place for a hive. I haven't been able to distinguish any scout bees. But I know they are out there looking for a new home for the swarm.

Some say that 20 of the swarm bees are scouts, i.e. those bees appointed to go out and look for a new home. Others say as many as 100 of the honeybees leave the swarm to scout out new digs! Who really knows?

I was reading that the scout bees have to determine whether the new location for the hive has accessibility to nectar and pollen. Is this new location easy to fly into and out of? Is it the right size for the number of bees in the swarm? Is the new hive defensible? I wonder how they determine who is a scout bee and who is not?

The scout bees might find that an existing hive is their best location. It is, obviously, occupied by other bees. What to do? The scouts have to determine whether their swarm is able to invade, battle and overcome the other bees. Then they would probably have to "persuade" their sisters to undertake this dangerous maneuver. If they lose the battle, they are so weak that they usually die out. If they win, they get the hive, the honey, the pollen AND the developing brood that they will tend and raise as their own. This is not the preferred method of finding a new home for the swarm, but it has been done.

Relationship of Bees left in the Hive to Bees in the Swarm---NONE!!

Even though the swarm is only c. 30 feet away from the hive, the hive bees come and go over the branch where they hang as though the swarm isn't even there and the bees who intermittently leave the swarm show no interest in their former home or housemates.

1:30 in the afternoon. They have been on the branch three hours now. The number of circling bees is down to 8-10. Perhaps they've determined it's a safe place and can withdraw some of the guards?

4:40 P.M. The swarm is gone! I can't believe I missed seeing them leave! About 4:15 I showed the swarm to Daniel and to Carrie and just now, less than a half hour later, I came out to see them and GONE.

Still about 12 bees are fussing around the branch where they stayed. Are they nectar gatherers who were away and came back too late? Is this the clean up crew? What is there to clean up? If they were away and came back too late, is the new hive location information in that area for them to find?

I'm so sorry I missed their leave-taking!

I sure didn't miss the swarming---all three of them! They hung on the branch from 10:30 to 4:30---6 hours. It's supposed to rain tonight and tomorrow. Wonder if they knew that and had to get into their new home and get settled?

Just think these swarms all had to find homes within miles of my home and their hive. I live in a small town in Ct. There are trees that bees could go into but they are usually old trees that have hollow spaces in them. The "bee gums" that people write about are old logs or trees. In the Middle Ages beekeepers kept their bees in straw skeps (woven baskets about the size of a small waste basket). In ancient Egypt and many other places around the Mediterranean, people kept bees and still do in long cylindrical tiles. In NYC several years ago they found millions of honeybees in

A Cosmos in my Kitchen

the walls of a big building. Maybe my swarm went into someone's home?

Wherever the swarm goes, the colony will begin to lay wax for brood comb and honeycomb. They will use the top part (roof) of whatever structure they choose as the support for the vertical, parallel wax hexacombs that they will begin to create. Think of the hexagonal combs as row houses. For the combs are homes for their young and for their food and provisions.

June 14, 1982

At 11:25 A.M. I heard a loud buzzing sound coming from the hive. Having heard this sound before, I have come to associate it with a sun break from bad weather. Soon after that loud sound, the sun came out.

Well, this time I timed it to see if they would be right or if I was right in my surmises. 11:25 announcement from hive, "Sun is coming out. Get ready to go outside." 11:40 The sun does, indeed, peek through fifteen minutes after that particular buzzing sound. I have noticed that there is appx. 15 minutes from the announcement---the buzz series---to the appearance of the sun.

So they know (do they?) ahead of time when the rain/cloud/gloominess will be dispelled by the sun and they announce it to the whole hive. The announcement is so loud I can hear it sitting in the kitchen.

June 15, 1982

Observation and Conjecture:

Some of the inner-hive dances could be announcements of the completion of tasks.

1. "Finished larva feeding."
2. "Finished capping."
3. "Finished cleaning the cell. It's ready for laying."
4. "Just examined this cell. It can't hold any more pollen, honey."
5. "Finished carrying out the dead brother." Etc.

The dances inside the hive, i.e. the body movements that are varied and have been by me repeatedly observed, occur both on the brood and the honeycomb.

12:50 at night. Steve and I just heard the beeps from the hive again. It is a long call which sounds like the blowing of a conch shell. Then after the long call, the Yeee(s) begin. The long call that I hadn't heard before reminds me of an ancient call to arms.

I think now there are several (at least two) bees doing these sounds. We have heard about ten (10) series of these so far---a long call followed by Yeee. One series had 40 Yeee. Another had 31 Yeee. Another 16. Several had 41 Yeee. Forty (40) and forty one (41) Yeee are frequent. There is one bee on this side of the hive doing it, but I can't determine which one. Remember, Sandra, the Yeee is pronounced "e" like in "me."

Another bee inside the hive often answers this call. The interior answer is not as audible and I have to put my ear to the vent hole to hear it distinctly.

A Cosmos in my Kitchen

What in the world does this series of sounds mean?

Here's another one just now. 41 Yeee. The answer was 10 Yeee.

Maybe the drones make the sound. They are big enough. No. It's those queens piping to each other.

Here's another: 39 answered by 12.

One on my side: 45 Yeee---Far-away answer---10 Yeee.

Forty four (44)---Answered by 9 Yeee.

Now they're calling back and forth rapidly. The one on my side, the very audible one, always beeps three to four times as many beeps as the one who answers on the other side of the hive. They're really going at it now!

The nonce the one stops, the other begins.

The one near me always does at least thirty (30) Yeee.

The one who answers has done as little as three (3) Yeee.

It's the one that is far in the hive that makes the conch-shell-like sound before the Yeee, not the one on this side. The one on the observation side just starts out with the Yeee.

Again concomitant with this "conversation of Yeee(s)," I observe that the workers in the hive have come alive with much more activity. They are stimulated in some way by this repartee. It is late at night and the hive is usually pretty quiet, but now there is an audible hum as a result of the piping.

I put the wood panel over the glass and the hive has been closed while I listen to this phenomenon. Now I'm taking the wood off and will be able to see inside. I'll see if this has any effect on the beeping.

Yes, it does. The one on this side is quiet now, but the one in the interior keeps on with a series of 7 or 8 Yeee. Obviously the one near me feels or sees the effect of the open hive and is lying low. The other probably doesn't know the hive is opened.

Wait. Now they are both Yeeeing at once. Not a call and reply, but both talking at once.

I do think it MAY be a drone sound. Intermittently I hear the low buzz of a drone between the beeps. However, the Yeee is higher than the low buzz of the drone. Obviously, I don't know.

They have been chattering like this for 30 minutes now minus the minute or two when the one on this side stopped right after the hive was opened. They go at it incessantly, aggressively. One would think they would tire or their little bodies would not be able to make that sound for such a long time. It has to be wind forced out of the abdomen and that takes energy! The sounds can be heard distinctly within 4 feet of the hive, but if I move back farther, they can't be heard. The calls do cause quite a stir among the other bees.

It's 1:30 in the morning so I don't think the beeps have to do with sun prognosis as when I first heard the sound days ago in the afternoon. My observations continually make me revise my conjectures!

A Cosmos in my Kitchen

Note inserted later: I found out, as I've indicated, that the queens were making those Yeee sounds. The experts designate that sound as "Ze-e-ep. Ze-e-ep." To me it sounds like "Yeee, Yeee." I don't hear the "z" sound or the "p" on the end.

But I'm not as trained as the experts. Also, there is argument among apiculturists (bee professionals) as to HOW the queen's piping sound is made. I posit that it is a sound FORCED UP FROM THE CONSTRICTION OF THE QUEEN'S ABDOMEN AND EMITED THROUGH HER PROBOSCIS (MOUTH). She squeezes her abdomen and forces the sound up through the thorax and out the mouth. It is a rhythmic constriction totally controlled by her and initiated in response to certain stimuli. The only other sound that bees make is a hum. The humming is caused by the flurry of their wings and not by a sound coming from their mouth. Some experts say the queen's piping sound is produced by the vibration of small plates at the wing bases. Other experts say she presses her thorax against a hexagon and the hexagon acts as a sounding board that radiates the sounds. I don't know. I think I'll go with my theory until the experts all agree. It's just so logical to me. By the way, Why does a bee hum? Answer: Because she doesn't know the words.

The hive is still congested. They'll have to swarm again. No sign of a new queen.

IDEA: Since I heard the first Yeee(s) three days before they swarmed, could that be the announcement of the birth of a new queen? Maybe the sound has something to do with the queen?

6/7--Heard Yeee sound

6/8--Rainy

6/9--Partly sunny

6/10-They swarmed.

Maybe, also, that's why I've never heard the sound previously. Maybe tonight another queen came out because I have looked in vain for the last five days and haven't seen one? I'll look well for a queen tomorrow and if I see one, I'll use this as a working hypothesis.

They're still Yeeeing tirelessly. It's been over 50 minutes. I'm going to bed. 2:50 A.M.

June 16, 1982

Believe it or not, it is 7:45 in the morning and they are still Yeeeing! The bees in the hive are going crazy! There is a high hum. They are outside already and running about within the hive.

8:20 A.M. 68 degrees. Sunny. The bees are swarming!! Maybe the Yeee is a pre-swarm sound---"Get ready, girls. We're leaving at sun up!" Maybe that is the queen, queens telling them to make the hive ready for departure?

They're swarming but not landing on anything. Just circling outside the hive and all on the inside of the window. Maybe they don't have a queen? I think the hive is filling up again.

They're swarming, i.e. behaving exactly like bees when they swarm, but they are not leaving. Is this swarm fever?

8:35 A.M. Thousands of them are in the window and filling the sill, the ones on the window sill are milling around rather than leaving with a purpose as they did when they swarmed the other times.

One bee, laden with pollen, is resting on the window. Maybe she is awaiting the cessation of activity around the bee gate before she goes in to deposit her gold.

I'm sure the swarm is aborted for whatever reason.

8:40 A.M. Still loads of excitement: bees, noise, activity, but no exiting.

1:00 in the afternoon. They're swarming again! These are the swarmingest bees! This time they're settling on the same evergreen. Each time on the same evergreen, but on a different branch or fork in the tree. It is 1:13 and most of the swarm is settled in a beard on the tree, but there is still a high pitch of activity in the hive. Loads of bees are still in a fever on the window even though they haven't chosen to join the swarm on the tree.

Note: I've seen pictures of people with a beard of bees on them. Some had thousands of bees covering their face like a real beard. Some had the beard of bees from their chin all the way down to their knees. One I saw (and put a picture of this up on the refrigerator in our Vermont house) was of a young man totally covered from head to toe with a beard of bees. The only features you could distinguish on him were his eyes, his nose and parts of his lips. In order for the beard to be on the person like that the queen has to be in there somewhere and the bees have to be totally engorged with honey so they are quiescent. Plus they'd have to be "coaxed" by someone onto that other person.

Maybe the reasons they couldn't get the swarm off the ground this morning when they were in swarm fever are:

1. The queen was too newborn.
2. The queen was not mated. Does she have to be mated to lead a swarm or can a virgin lead out a swarm? If she has to be mated, maybe they were in swarm fever as she flew out this morning to be mated?!
3. It was too early in the morning to swarm. The others all swarmed after 10:00 A.M.
4. They hadn't completed whatever preliminary work needed to be done before swarming:
    a. Stock up on honey.
    b. Find the right branch to swarm on.
    c. Complete hive work.
    d. Were in a fever, but were not properly organized.
    e. ?????

I'd like to EXPLORE THE POSSIBILITY that they were in that swarm fever because A VIRGIN QUEEN WAS EXITING TO BE MATED. It's a fascinating conjecture that the whole hive gets excited when a virgin leaves and the drones from her hive as well as any other nearby drones take off after her.

Maybe, yes, that is what happened this morning. A virgin queen left the hive. It is a sunny day, good conditions for a mating. None have ever seen the mating of the queen and drone in the wild, but some have claimed to hear a crack high in the air. I wish I could have seen her take to the air and seen the drones in the hive, their raison d'etre moment having arrived, take off after her. She is bigger than they and fast. She soars higher and higher and they push the air with their big, powerful wings in pursuit. Hundreds of them from my hive, and others who SMELL her scent from other wild hives nearby, wing in pursuit. I've read that the drones appear as a comet-like tangle of black spots in the air. Higher and higher she flies. Only one will catch this pristine beauty, this virgin queen, this matchless monarch. And he leaves the others behind and keeps his multi-faceted eyes fastened on the abdomen of the prize ahead. And he catches her and in mid-air mounts her and grasps her with his powerful legs and enters her and comes. The crack must be at that moment for the force of his orgasm tears off his penis (endophalllus) and part of his abdomen and he falls, fatally mutilated, to the earth. She must fly around for a while and then begin her descent back into the hive. And she enters my hive with his penis and abdomen parts trailing her as trophy, as proof that she is mated. (Remember the Elysian mysteries' SHOW from the Mary Renault book The King Must Die? The king had to take the soiled sheet and show the people he and his bride had had intercourse.) And the workers remove the queen's SHOW and accept her now as their queen. His moment of ecstasy---for there must be that ---will live on for many generations of bees. As long as she lays, he lives. His strength, for he was the strongest. His endurance, for he outlasted the others. His daring, for he did. Driven to a doomed mating by determined destiny. And she, no longer a pretender to the throne, but by his death a fecund ruler now, loaded for life with all the fertilized eggs many generations of the hive will need. And she begins to lay and lays and lays and lays. Fulfilling her fecund fate as he did his. How beautiful it all is.

And did she mate this morning and did she hours later lead them out? When she returned, she was no longer a virgin lady in waiting. She was Queen. I haven't read anything about feverish in-hive excitement when the queen goes on her mating flight. But maybe that is what I observed?

The hive is still very congested and that is really a sizable swarm out there.

This swarm is twice or one third again as large as the last swarm.

I can see the brood comb now. There is some capped brood on the top and a few scattered throughout, but most of the cells are empty, pollen-packed or filled with nectar.

7:10 in the evening. It has been thundering, lightning and raining furiously for 10 minutes now. The bees are hanging there in their cluster in all this. God bless them.

7:25 P.M. I couldn't stand it. I had to go out and check them in all this storming to make sure they were okay. I climbed up the tree and got within a foot or two of them. Usually the swarm is compact, but there is movement, fluidity, activity, with bees coming in and out of the swarm.

This Storm Swarm is extremely tight, totally immobile. They have formed themselves into a GIANT FISH SCALE. As a result of the formation, the water just runs off of the swarm and drops to the ground. It is quite a beautiful, elegant formation! It is really pouring. Only the outside, fish scale bees, are getting drenched and they appear none the worse for wear. Also of help, the tail of the swarm is shaped like an amphora, one of those ancient oval alabaster wine casks that must be set in a holder. That streamlined bottom also forces the water to run off at the upper parts of the swarm and not congregate at the lower end. Good swarm formation, bees!

Note inserted much later: When we (Steve and I, Kathy and husband Walt, Blake and Jesse) visited Crete in 1996 on our 25th wedding anniversary, we saw ancient amphorae, cracked, chipped and broken, in storehouses in the Palace at Knossos. They were the exact shape of the bees' Storm Swarm. The Palace of Minos is the legendary home of the monster half-man/half-bull Minotaur. Mary Renault's The Bull From The Sea is set on the island of Crete in the time of the Minoan civilization (c. 1450 B.C.). It retells the myth (?) of Theseus and Ariadne and King Minos and the Minotaur. One thing I love about Renault's books is that every cup a character uses, every dress a woman wears, every ring on every finger is authentic. She takes all these extras from ancient vases, bas-reliefs, mosaics, etc. When we were at Knossos, the authorities had replaced all the ancient mosaics in the Palace with terrible fake ones. I was VERY disappointed with the ruins, especially since I had studied the Minoan civilization since I was 17. It was around that time that Michael Ventris had succeeded in deciphering the Minoan Linear B code of writing. I was (still am) interested in all things archaeological and this was big news to me. Of course, it was HUGE news to archaeologists. The Minoan scribbles had been thought undecipherable. Ventris had been dedicated to deciphering Linear B since he was 16! In 1954 his young and innocent face was splashed all over the world. He was dead 2 years later at the age of 34. But his name and achievement live on as do the names Theseus, Ariadne, King Minos, the Minotaur. And those ancient amphorae, emptied of their elixir, have a certain fragile immortality, I guess.

Sandra Sweeny Silver

June 17, 1982

Looked at the swarm at 2:45 in the afternoon and it was still there. Looked now at 3:10 and it is gone! I always miss their leaving! Darn.

It rained very hard all last night and they hung in there. It was cloudy this morning, but this afternoon, it got sunny and was in the high 70's.

After the exit of two swarms now, the hive looks good, normally congested, very USED. Just think, almost three generations (or more) have lived in this city. For them---average life span is c. 4-6 weeks---a month is a lifetime. Time is relative, for sure.

There are probably about 10-20,000 bees in there now.

June 19, 1982

1:30 A.M. When the last swarm left, it was such a big one that they engorged all the nectar from the honeycombs. Bees who will leave with the swarm all fill up on nectar before leaving. Filled with honey, they are peaceful, non-stinging. Hence a "Sweet Swarm." That was three days ago and still the honeycombs are empty except for 15-20 cells in the upper left hand corner.

Most of the brood has hatched. Very few capped ones left. So they're all ready for a new batch, a new generation. Get busy, queen.

I haven't seen any queen since I put the doomed, clipped queen in after the second aborted swarm on 6/2/82. In these 17 days, they killed the useless, clipped queen and built a viable queen (takes c. 16 days to build). She was mated? They swarmed on 6/10/82. Then they built another queen or two and swarmed three days ago. They've been very busy trying to get the population in the hive down-sized so they can work unimpeded.

Just since the last swarm, the population had increased. Most of the capped brood had emerged. Will they spin off another swarm in order to get the population down to 8-10,000? Unlike humans, they will NOT live for long in an overly congested city. It's just not "civilized." And my bees are, in their instinctual way, very civilized.

I think, 5,000-8,000 would be an ideal population for this observation hive with only three brood combs and three honeycombs.

I continue to feed them sugar water. They consume a half a bottle every 18 hours.

Yesterday I was in the yard reading by the large clump of Allium senensis. I heard a furious buzzing sound and thought one of my bees was stuck in the ornamental onion. I casually passed my hand over the flat, spiky leaves and went back to reading. I thought that might release her. After c. 5 minutes, the buzzing was as furious as ever, so I began to search for her and found---him and her---mating! Not a pair of my honeybees, of course, but a big, furry bumblebee and a queen, bigger and furrier than he.

I called Blake and Bobby and we watched.

Number I. The act took c. 20 minutes (as long as we watched anyway).

Number 2. The female braced herself against the flat, broad leaves and except for several attempts at re-positioning herself, she was immobile, totally passive.

Number 3. The mating took place near the ground at the bottom of the clump of allium.

Number 4. The male made an incredible amount of noise and was very physically active. She---noiseless and passive. It was his noise that originally attracted me.

Number 5. They were locked---he on top of her, buzzing, moving and really "drilling" her.

Number 6. I moved leaves here and there around them searching for angles of observation and information. They were aware that I was there, I know, but gave me no heed. He was going to finish---no matter what.

Number 7. When they were finished, he withdrew. She steadied herself, and she really had to steady herself, on the leaf. He fell to the ground for several seconds. Then IN CONCERT they flew up the clump and into the air. She flew away to the left. He flew to the right. Gone. Blake and Bobby were as fascinated by this as I was.

Since we can't observe the honeybee's mating because of the altitude, I thought these small eyewitness observations in at least the same genre, Apis, probably hold true for the mating of Apis mellifera (minus male death). Also, I'm thankful the Lord allowed us to see this personal moment.

Summing up:

1. It's not a short mating. It takes 15-20 minutes. Could be longer because it was going on for a while before I became interested.
2. She's passive and receiving him.
3. He mounts her and is active, noisy, and moves around on her back a lot.
4. When it's over, there's a moment of rest and reorientation for both of them. I don't know whether he goes on to die because I don't know that genre of bees. But for her, when they rise in the air together, it's a spin and a return to her duty and destiny.

Note inserted later: No, the male bumblebee does not die after mating.

Re: Queens and Supercedure Queen Cells

Blake observed the clipped queen in the first weeks of the hive come upon a cell in the lower right hand corner of the brood comb. He said, "Mom, she tore that cell apart with her mouth." I've since read that a queen will do that to other queen cells to keep new queens from emerging to challenge her hegemony. He said she tore the cap off and was determined as if on a search and destroy mission.

June 20, 1982

It's Steve's Birthday today! Plus it is Father's Day. Praise God for our own King Bee! The way he works, he's no drone!

The brood comb is now mahogany black on the inside and the hexagon rims are the color of the bees--- gold/brown. They've gone through at least 2-3 sets of brood and they show it. The honeycombs are still white and have never had the chance to be capped because there have been so many bees and they are in constant need of nectar. Plus two swarms have left laden with nectar, so the honeycombs are still pretty white.

June 21, 1982

This morning at 11:00 I was out examining the heath, heather and thyme garden in front of the carriage house. It is bordered by two giant ilexes. I really know nothing about bushes and had always appreciated these old bushes for their evergreens in winter bouquets. I had never examined them closely. I simply cut their proffered greenery.

I noticed that hundreds of my little bees were hovering around the ilex. Why? They seemed to be exiting the hive that is about 80 feet away and going into the bush. I turned over one of the small branches---tens of thousands of TINY greenish white flowers were blooming!! One would NEVER know they are there because all you can see when you look at the bush are the perfect, waxy, little leaves. But hidden underneath is a world and wealth of perfection. (I'm sure these little flowers turn into the black berries that are under the leaves in the winter.) What were my little ones getting from these little ones? It was nectar because none of them had any pollen in their baskets.

The honeybee loves, prefers, seeks out the tiny, tiny flowers which no one else cares about or even sees. But she with her thousands of facets sees them and cares about them and carries their nectar into the hive to bring her sisters and us our honey, our gold.

Just think, the honeybee has maybe 4,000 facets in each eye! And the shape of EACH facet is a HEXAGON!!!

A Cosmos in my Kitchen

And each of those facets in each eye has its own lens!!! We humans have one lens in each eye. Imagine, it's impossible of course, what the honeybee sees! About 8,000 lenses at work at all times. Scientists can only speculate the view of the world they have. They call their view "pixilated."

It is interesting, however, that each facet is hexagonal and they do build hexagonal structures!

I rarely see my bees on the large or medium-sized showy flowers. Logic would dictate they would prefer big, fat nectar fields to teeny, sparse foraging fields. No, they are into William Blake's "Minute Particulars... World in a Grain of Sand" cosmos. And they have called my attention more and more to the microcosm, the micro, the m.

"Pixilated" also means "drunk." "Slightly eccentric or mentally disordered" is how the dictionary defines "pixilated." That qualifies. When women are drunk, people use such soft words and phrases to describe their state: "she's in her cups," "under the weather," "a little tipsy," "schnockered." There is something very embarrassing about a woman drunk. Drunk men, on the other hand, are described in bold and ugly ways: "he's shit-faced," "pissed," "hammered," "locked and loaded." Minces no words. I like the similes for drunk: "he's high as a kite," "drunk as a skunk," "boiled as a hoot owl." There's the edgy, trendy ways to describe drunks: "totally zoned," "spanked," "bent," "fried," and "stoned." Then there's the erudite descriptions of over-imbibing: "inebriated," "crapulous," "besotted" and, of course, "pixilated." I don't drink which makes me a "square," "antediluvian," "fuddy-duddy," "stick-in-the-mud," "fossil" and a "mossback." It's 2:12 A.M. This teetotaler is going to bed.

June 22, 1982

Have to finish up on this "teetotaler" word. The etymology of the word "teetotaler" is disputed. Some say it came from those who abstain from intoxicating beverages and only drank "tea." Totally tea---nothing else. Others say the "tee" is just a repetition of the "t" in "total." That would mean a teetotaler was an abstainer with a double "t." That double "t" emphasizes the absoluteness of not drinking alcohol.

12:20 midday. Re: the nectar gathering on the ilex bushes.

When I concentrated on one bee, I counted 56 flowers visited before returning to the hive. She sipped in about a one square foot area. She didn't fly all over, stayed compact, concentrated in one area. She went methodically from one branch to the adjoining or nearby branch. Seems to me their nectar tongues are always extended in those situations of extreme abundance, only retracted sporadically.

6:23 P.M. The bees are still foraging on the ilex, so those flowers must give nectar all day long rather than only for several hours like some flowers.

It is true, though, that "the early bird gets the worm." Most flowers secrete the most nectar in the morning. There are some flowers that depend on nocturnal pollinators (moths and bats). These plants secrete the most nectar at night. Some flowers only yield nectar for several hours a day and the limited flow attracts a certain type of insect. That insect is its prime cross pollinator. I wish I could remember what flowers they were.

Hummingbirds are "nectarivorous" birds. No morning worms for those early birds---just the sweet treat! And the nectar has to come principally from flowers in the red/orange-red spectrum. Otherwise, the long beak of that dainty bird will not be pointed in her direction. The hummingbird is the main pollinator for over 60 families of plants! Watch her hover over a red flower with her little wings whirring at 50 times per second. This bird is teeny. The smallest hummingbird is only 2.2 inches long and weighs just .067 ounces! The average ruby throated hummingbird that we have here in Ct. is 3.5 inches long and weighs a whopping .14 ounces! That's only amazing when you know that the hummingbird has to migrate south in September and fly 2,000 miles to Central America!! Just think. That little 3 inch bird weighing less than an ounce flies at 25 mph more than 1,000 miles down south and then flies non-stop for 20 hours across the Gulf of Mexico (a distance of 525 miles) and then has to fly another 1,000 miles to get home to Central America. I'm dumbfounded. "Found to be mute. Struck dumb. Mute." I can't write any more. The wonder and beauty and awe of it all! God.

1:10 A.M. Still no sign of a queen even though I look all the time.

The bees are bringing in propolis, I think. Propolis is a brown, gummy substance the bees use to seal up certain places in the hive. It's appearing now in greater abundance than before. They are using it to seal around the glass and honeycomb. They have always stuffed things into this space. Mainly white grain (?) that looks like sand and now there are some legs and antennae stuffed in there. Are they sealing these things away for hygiene's sake? Why didn't they take the body parts outside?

the bees at the door of the hive are aerating the hive or evacuating so that it can be cleaned or maybe it's too hot in there or.....

They have taken one-tenth bottle of sugar water in more than one and a half days. Nectar must be plentiful. This is the least they have ever consumed. Also, I'm finding it hard to get the nipple of the bottle in and out of the hive. Bet they are propolising around that hole, too. Propolis does have antibiotic properties. They don't want those kitchen germs coming in!

No sign of larvae on the brood combs that I can ascertain through the mass of bees. It's definitely crowded. But there are nectar and sufficient pollen.

There is a preponderance of nectar stores over pollen stores. I reason they must consume more pollen than nectar, more need for pollen at this brood-feeding time than for nectar. OR there is a greater ratio of nectar gatherers to pollen gatherers, OR NECTAR is the thing---honey seems to indicate this is true, but all could be true. And more I don't know.

Nectar is definitely the thing from an overall standpoint. Pollen is important for feeding the next generation. But it is one of the by-products of the hive---a crucial one. Propolis and wax are important by-products, also. But NECTAR IS KING! A normal colony of honeybees brings in about 528 lbs. of nectar a year! After the evaporation and other processes, that yields about 132 lbs. of honey a year.

528 lbs. of nectar carried in ounce by ounce by those little forager honeybees!! It boggles my mind!

June 30, 1982

Sometimes when I open the hive, they begin a flurry of activity and agitation in a certain pocket---usually the bottom middle. They wiggle their bodies sounding an alarm. They face the glass running up and down. They circle back and forth. The rest of the hive remains normal, unaffected. There's just that semi-circle of activity when I open the hive.

Is that spot the beginning of brood on this side and are they protecting it? I experiment and close the hive for several minutes. I open it. Still quiet. Then a flurry of activity. Close it. Open. Quiet. Flurry of activity. So it IS in response to the opening of the hive.

NO QUESTION I HAVE IDENTIFIED AN IN-HIVE POLLEN DANCE! With full pollen baskets they shake the abdomen violently back and forth sideways. The head and thorax don't move---only the abdomen with the full baskets. They wander around for a long time, often in circles, doing the Pollen Dance.

The shaking itself lasts for as long as it takes me to verbalize out loud: "One One Thousand." Wonder why I've never read about the Pollen Dance WITHIN the hive? It's certainly the most easily identifiable one in the hive. There are five (5) pollen dancers right now in that area of high activity in the bottom middle, but I've seen more of them depositing their load on the brood comb. Pollen for new brood? Hope so.

Talk about the Pollen Dance. No wonder they're dancing! I was just looking at magnified pollen grains. Wow! I can't believe how beautiful each pollen grain (should say "mote of flour") is! A grain of corn pollen is smooth, oval-shaped with several side slits. This grain of Himalayan blue poppy is round with little pointy pocks all over. The cherry pollen is gorgeous---a lumpy round with thousands of swirls. The hellebore grain is round, elongated and looks like it is encased in loose macramé. The round bull thistle pollen looks something like its flower---very pointy and bristly. Because of the degree of magnification required to photograph each grain of pollen, each grain looks like a planet with the universe behind it. To think that there is in each plant and here in my hive such a cosmos of wonder! Each grain of sand and each snowflake is different from any other grain of sand and any other snowflake. Now each grain of pollen is unique not only to its plant but in and of itself. Each person, though a human being, is different in mind, body and spirit from any other human being who ever has been, is or will be. Each pollen grain from an apple tree is distinctly identifiable as from an apple tree. HOWEVER, each microscopic grain of pollen from that tree has its own unique identification. This is mind-boggling to me, Lord! This infinity hidden within the finite world! William Blake's mind was blown by the infinity within a grain of sand. My mind is being blown by the infinity hidden in each grain of pollen!!

July 4, 1982

Independence Day. I'm going to have a pool party later on and then a barbeque for 100 people.

11:30 in the morning. We've spotted the new queen! She's not quite as big as the original clipped queen, but she's still bigger than the workers and easy to spot. I opened the hive and there she was on the top of the brood comb with space around her, highlighted. I called Mother to come over (who had never seen a queen before) and we watched her for 45 minutes. She wandered the comb ferreting out ready cells and then inserted her abdomen and laid. Appx. time of each laying was 21 seconds. From her wanderings, her speed, the amount of laying, the number of attendants (6), I feel she is:

1. Just getting into the saddle.
2. Not as awesome, charismatic as the other queen (to these workers).
3. Maybe not as potent. I'll have to see from the brood pattern.
4. Size is smaller. Could mean not as genetically well-bred?

This queen does an interesting thing when she lays---not every time, but occasionally. She inserts her abdomen. Her head is facing toward the TOP of the hive. As soon as she gets really into the cell, she turns her whole body about a third of a circle. She rotates her abdomen and ends up facing DOWNWARD to lay rather than maintaining her initial position as did the first queen. The clipped queen never, and I observed her hundreds of times, did this queer rotation.

The brood comb sparkles with nectar and is punctuated with packed pollen. (Alliteration. The worst alliterative phrase I ever tried was in graduate school in a paper on "The Faerie Queene" by Spenser. I said, "Fanciful, faerie-filled fields.")

Some of the nectar is ripe and capped on the top right and left hand corners of the brood comb.

Today I was walking past the evergreens on the way to the tennis court to check the wildflowers on the bank. I heard a ticking sound---tick, tick, tick. What was it? I began looking in the evergreens. On a branch was a strange looking structure. It was round and smooth and looked like it had been thrown on a potter's wheel. It was grey-white and had been woven into the fabric of the evergreen branch and needles. It was a nest of some sort. I saw insects flying in and out of the bottom of it. I moved closer for a better look at the bottom. The most amazing structure! The bottom looked like a perfectly finished pot with a rim for standing. It's hard to describe how perfect and symmetrical and beautiful it was. My mother thinks it's a hornet's nest. It probably is. I've never seen one.

Steve said we have to get rid of it. My mother agrees. But I don't know. They could be dangerous to Jesse, 6, of course. I'll find out and if they are, they will have to go.

Note inserted later: It was, indeed, dangerous. Steve and Blake got rid of the hornets with a spray. I insisted on keeping that gorgeous structure the white-faced hornets had made. I cut it off with the evergreen branch attached to it. It was integrally woven into the branch and needles. I examined it. Fascinating. I, also, took it to Jesse's 1st grade class and showed it to them.

My bees are still feeding on the ilex. So far they've been most devoted to those bushes and the ajuga. The big bumblebees love the big flowers. My July-blooming rhododendrons are currently their favorites. My honeybees like the white clover that punctuates the grass, too. They are not partial to the big, fat, pink clovers that I allow to invade my garden because they are so nice.

The honeybees are obsessed with our swimming pool. I bought a small, cement shell and have kept it filled with water on the ground right outside their hive. But no---they like the pool. I realize the pool is more visible from the air than the tiny bee bath I made. However, many of them have been drowned in the pool. They land on the cement near the water, attempt to load up, fall in and can't get out. All the family has rescued dozens of them. Cries of, "Quick, one of the bees is drowning!" are common around here now. I've even attempted to warn them through the hive, but they don't listen.

August 1, 1982

It's my dear Jesse's 7th Birthday. I praise You, Lord, for the gift of this sweet one. Bless him and keep him close under Your Wings always!

Almost a month since I've been able to record---too much entertaining around the pool. Today for Jesse's Birthday I'm having a party with 30 teenagers and six seven year olds. Jesse, ten years younger than Blake, twelve younger than Kathy, has lots of teenage friends! They love to come to his parties and play Ring Around the Rosie, Pin the Tail on the Donkey and Drop Clothes Pins in a Bottle. They get into it with the little ones.

This new queen is great. The brood pattern is every bit as full and regular as was her mother's. So many are there and hatching. They must swarm sometime soon.

They have propolised every part of the hive where there could remotely be any outside air coming in. They keep bringing in the propolis. I've seen them packing and working with it but have never seen it carried in. How exactly is that done? They constantly pack and pick at it with their forelegs and mandibles.

They are, also, waxing over some places: the half-inch spaces on the left and right ends of the honeycomb between the comb wood and the hive wall. The wax sealing is wavy. Probably because a wavy line breaks less easily than would a straight wall of wax?

ALL the honey in the upper comb is capped, finally. They have consumed NO sugar water for a month now even though it's there at ready. I'm sure they have put propolis over the little nozzle hole and the nozzle because I can't get it out. Will have to work it out some way come cold weather.

August 25, 1982

Like my bees, I've been working all summer---laying in food and cleaning up my hive for the hordes of people that keep coming. Occasionally I wish I had a few workers to help me with these two houses, pool, tennis court and 3 acres.

I do have such great joy. Several nights ago Jesse, Blake and I spread a blanket in the yard and looked up at the stars. I used to do this all the time in the summer when I was a kid. It was a perfect night. Dark sky. Bright stars. The Milky Way was very visible and its cloudy mass of stars always boggles me. An infinity of stars JUST in that one spot of the heavens. The ancients used to call the Milky Way the "road to the gods." They felt the same awe as the boys and I do. King David spent a lot of time hiding in the desert from the jealous insanity of King Saul. He, who knew that there was only one God, often laid out under the close, black desert sky. He has described for all of us who believe the EXTREME humbling one feels when contemplating the universe of night stars: "When I consider Your heavens, the work of Your fingers. The moon and stars which You have set in place. What is man that You are mindful of him?" (Psalm 8:3,4) Yes, why does the Creator even NOTICE let alone LOVE me? I'm not even a mote of dust in the great

A Cosmos in my Kitchen

universe! I'm nothing! But every thing in the universe says You, as Creator, care about Your handiwork! And every thing SHOULD lead people on this little planet to You. "The heavens declare the glory of God. The skies proclaim the work of His Hands!" (Ps. 19:1) Lying on the blanket, in the dark, in the middle of my two boys, pointing here and there, talking softly, being silent before the great extravaganza---I'm really in heaven!

They are obviously not going to swarm even though there is an over-abundance of bees. They see the leaves beginning to fall into the pool. They feel the nights getting chilly. They know, as I do, that the short sun season is over and the earth is beginning to crack.

Re: FANNERS There are always bees fanning on this side of the hive. They are positioned either on the comb with their wings and back to the glass or on the glass with their wings and back to the comb. I've never seen them fan on the honeycomb, only on the brood comb. NO! Right as I wrote that, I saw one fanning on the honeycomb. So I stand corrected, as I must with all my primitive observations! But most of the fanners are on the brood combs. There are at this time of year about 4-6 fanners on the combs at all times.

We know why they fan. It's because the temperature in the hive is either too hot and they have to air-condition the hive or it is too cold and they have to generate heat. The ideal hive temperature for them is 91-96 degrees F. It is August and I'm sure with them packed into that small city it is too hot for them. So the fanners are fanning to cool down the hive.

What I want to know is WHO DECIDES WHO SHOULD FAN? I have that same question as to who (or what genetically) decides who should be guards or who should be in the queen's retinue or who should bring in pollen, nectar, propolis. I'm sure there are multiple hypotheses in the bee world regarding this question of division of labor, but can we really ever know for sure?

Honeycombs are still totally capped---so gorgeous, undulating white with a golden promise.

Almost all the brood has hatched. Will she lay again knowing the time is short?

Most of the brood comb now used for honey and pollen stores. Some of the honey is capped on top of the brood comb, as usual.

Even though it's 6:40 P.M. and cold, 67 degrees, they are still bringing in pollen. Look. This particular bee does not know where she will deposit this load. She goes around assiduously, searching to deposit her load in just the right place, stopping now and again to do her "I've Got Pollen" dance.

With no developing brood, there is much more space in the hive. They move easily around. In places they can huddle three-four bee deep between the glass and the comb.

These bees have revolutionized my reading life. I used to get old <u>National Geographics</u> from Thrift Shops for the archaeology articles. I still do, but I now forage for issues with articles on insects. Also, out by the pool, I have watched the mud wasps deposit their eggs in the sand between the blue stones. Other years I would be afraid and run from them, shoo them away from me, from the children. Now I get down on the hot blue stones and get as close as she will let me. Watch her back legs, better than a bulldozer, flick away substantial piles of sand to go deep down where her progeny will be safe over the cold winter ahead, where no one can

A Cosmos in my Kitchen

disturb their development. She dive bombs me if I get too close and is generally threatening, but I persevere and she vigilantly tolerates me.

Years ago I had an ant colony in a plastic ant "home." This was in Pittsburgh and I did stay up late into the night observing them in their white sand world. I marveled at the cantilevered paths they made. When I went to F.L. Wright's Falling Water, the Kaufmann Dept. Store family home, I remember the guide said Wright was influenced in his architectural structures by the cantilevered lines the ants make. They are very strong, can withstand stress, etc.

Then about five years ago I got another ant colony. This structure was made of wood and had beautiful variegated sand in it. The initial ants died, so I got some ants from outside and dumped them in. They made feeble attempts at feeling at home, but they died, too. I was fascinated by the look into a microcosm, by the order there, by my ignorance of this whole universe around me.

For the bulk of my years I have explored the universe within---the mind and the spirit. My hunger for those potentials will, I'm sure, never abate, but I am belatedly becoming hungry for the outside. And that outside seems to be focusing on the world of nature. Gardens are now of interest to me. I have dug twelve of them with my little shovel and have shoved hundreds of perennials into my dark and fertile earth. I watch them as I watch my bees, continuously and with excited interest. I guess you can turn into a different person after 40. I had always thought I was pretty well set in the interior intellectual, spiritual world and would ride that wonderful highway into eternity. That was not to be. Now it's the little worlds all around me. The world outside me. I just praise You, Lord, for giving me all these interests. They have made my life so meaning-full for me. Each day almost has been one of anticipation. Sometimes at night when I go to bed, I can hardly wait to wake up in the morning and thus I can't get to sleep! I know it's You, Lord.

Well, here's the queen again. Every time and I mean every time I have opened the hive for any length of time, she appears. This woman is quite the prima donna, the guardian of her people. She didn't lay, just sauntered from the left side to the middle right and went away after sipping nectar from one of the cells. She's not at all shy.

There are hundreds of dead bees between the window screen and the storm window on the upper part of the window. I don't know whether they have gone there to die, a huge bee graveyard (sharks and elephants go to a certain place to die) or whether they are placed there after death by the workers. I have never personally observed this new phenomenon. Astute observer Blake has seen them go there, flutter and die. They can certainly get out of that place, so they don't wander in and become trapped. The weight of the bottom layer of dead bees prevents the others from falling into the sill below.

Two to three months ago there were loads of dead bees on the sill at the entrance to the hive, but they are gone now. The sill is clean as a whistle.

In the four months I've had my bees, only one person---of the hundreds who have observed them and the hundreds who have been entertained on these three acres---has been stung. Only my little Jesse accidentally after that swarm bee got into the kitchen. (6/2/82)

The reactions to my hive range from Andrea Pritchett's, "How hideous!" to "How fascinating!" The "hideouses" are soon converted to fascination once I explain to them exactly WHAT they are seeing.

September 11, 1982

With my 4x magnification glass, I look deeply into the cells. They shine like waxed mahogany. Truly stunning, perfect, eight-sided wombs to receive the larvae and the life-bloods of the hive: nectar and pollen.

And, by the way, beeswax is a prime ingredient in lots of furniture wax! I've used it and it sure does make the wood shine!

The whole half of the top of the brood comb is capped honey. The rest---gleaming, empty, mahogany with occasional dots of golden pollen for striking color.

They are busy and humming today. It is gorgeous, cloudless, in the high 80's. They are redeeming the time and so, hopefully, am I.

There is a huge 100-year-old hydrangea tree near the tennis court. Bees of all species love it for its pollen. The pollen my bees gather from this old war horse is white. It's the only white pollen I've seen them bring in the hive. Mostly the pollen has been various hues of gold. Of course, there was the blue pollen from the scilla when I originally hived them.

The bees are "friendly" with us outside in the yard. I talk to them and pet them. I find them quite gentle souls with a divine compulsion.

September 17, 1982

12:40 in the afternoon. Just saw something I've never seen before. The bees were "rioting" on the window pane---singing, humming, hurrying to and fro, absorbing the strong sun. One bee was on the top sill between the two windows. She appeared to be grooming herself, proboscis semi-extended and licking her forelegs. The other legs were rubbing against her abdomen. Many (12-14) bees tried to or accidentally got near her and she "snapped" at all of them!

E.g. One bee bumped her and she turned around and batted her with her forelegs. Another bee was frolicking near her and she went after her with her mouth and chased her away.

In general she was "testy." She clearly wanted to be alone! She wanted and guarded her own space! I've NEVER seen bees bothered by other bees in such a "human" way. Togetherness is almost the definition of the honeybee. They tumble, fumble, crawl all over and hang on top of each other all the time with such abandon. Personal privacy is not in their genes. Yet this one lady just wanted to be by herself to do whatever it was she was doing. MOST UNUSUAL.

I notice today there are two capped brood cells in the lower right of the comb. I saw (the bees are covering the area) 10-20 larvae in cells. But the rest of the cells are empty of brood.

September 19, 1982

There's a lot of moisture on the brood comb. The honeycomb, however, is moisture-free. It's rainy, cold, 52 degrees outside. There has been moisture on the

A Cosmos in my Kitchen

brood comb recently even on sunny days. Hope it's not a problem.

Interesting note: Every once in a while "stuff" is on the OUTSIDE of the observation glass. I mean that on MY side of the glass, stuff appears. The bees cannot get to my side and have propolised up every possible air pocket. Where is this stuff coming from? How does this stuff, which I know is from the INSIDE of the hive, get to the OUTSIDE on the glass?? This phenomenon has occurred from the beginning days of the hive until now.

Just now I picked off a bead of amber-colored stuff---looks exactly like the propolis they've been bringing in. If I didn't know better, I'd suspect osmosis!! If I run my hand over the glass here, it is bumpy---little beads of this and that.

There has to be a RATIONAL explanation for this. How can stuff from the inside of the hive get to the outside of the hive when there is no possible exit point?

IDEA: Perhaps when they fan in order to condense moisture from the nectar and propolis, some essence of the nectar and propolis is fanned off and escapes through the minute edges of the glass and because it can't go "outside," it hardens on the nearest thing, the glass, with the removable wooden panel up against it. This stuff must then appear only when the wooden panel is up against the glass. That's maybe the reason for these constant hardened beads of stuff.

I just picked a bunch of it off and tasted it. I thought it would be spicy like the propolis, but it has no taste. It is waxy, chewy. Maybe then it is residue of wax production? If so, I'd have to come up with a more grandiose explanation. Then the question would be: How does residue from the bee's wax glands get out of the bees or out of the combs and suddenly appear on the other side of the glass? This stuff is, however, not white, but amber and CLEAR-colored. Used wax???? I really have no explanation.

This particular "stuff" is amber colored. I have a piece of amber in my jewelry box. It's yellow-reddish and is considered, I guess, a gem. But amber is really PETRIFIED resin from eons-old trees. Resin is the sticky stuff that oozes out of trees---especially from my many pine trees. It's hard to get off your hands. If an insect gets stuck in it, that's it. Forever. That's why we've found so many insects from eons ago in pieces of amber. We know, for instance, that the honeybee hasn't changed over time. (Neither has the cockroach.) The honeybee has been found totally intact in amber and she looks exactly like my Italian bee looks! They have, also, found a fossil of a bee imbedded in sandstone. Same bee as mine. I do believe in changes/evolutions WITHIN species and in the extinction of certain species. Scientists told us that the coelacanth fish had been extinct since the Cretaceous Period over 100 million years ago. But they were wrong. One was found alive and swimming in 1938 and another jumped onto Thor Heyerdahl's raft, Kon Tiki, in 1947. Since then coelacanths have been popping up all over the place. Not extinct after all. I'll not believe in Darwinian evolution of man from an ape-like until they find a man/woman/child that looks SIGNIFICANTLY different than me totally preserved in a peat bog! It's unscientific, totally speculative to postulate how ancient man looked in toto from just a scintilla of a jawbone or a sliver from an ulna!!

Man has been finding insects preserved in amber forever! The Roman writer of epigrams Martial (38-104 A.D.) wrote this apt memorial to a bee in amber 2,000 years ago:

"The bee enclosed and through the amber shown,
Seems buried in the juice, which was his own.
So honored was a life in labor spent:
Such might he wish to have his monument."

I just love this other epigram from Martial! In Latin it is:
"Non amo te, Sabidi, nec possum dicere quare;
Hoc tantum posso dicere, non amo te."

Most people know the translated poem:
"I do not love thee, Doctor Fell.
The reason why, I cannot tell;
But this I know, and know full well,
I do not love thee, Doctor Fell."

But very few know that this catchy and whimsical translation was done by a ne'er do well Oxford college student in England c. 1680. Tom Brown had done something bad enough to get him expelled from Oxford. The authorities sent Tom to Dr. John Fell, dean of Christ Church in Oxford. Tom's fate was in the good Dr.'s hands. Dr. Fell gave him the above epigram from Martial and asked him to translate it for him. Tom was excellent in Latin and had a great sense of humor. He made the Roman "Sabidius" in the poem into Dr. Fell. His on the spot brilliance won him a stay of expulsion. In addition to that, Brown's "Dr. Fell" doggerel has delighted English-speaking people for hundreds of years. (A loose translation of Martial's Latin epigram would go something like this: "I don't like you, Sabidius. I can't say why. But this I can say, I don't like you.")

October 1, 1982

The nectar flow is drying up. But Nature still provides for the sweet bees. The purple, pink and white wild asters thrust their sturdy stems up all over the property. My bees and other bees and insects are getting nectar from them. Right next to a particularly large stand of the wild white asters are the cosmos which I have planted all along the front of my vegetable garden. Now talk about flowers! These old-fashioned beauties will NEVER go out of style with me! They are so profuse, so reliable, so impressive in groupings and their stems are sometimes as big around as my ankle. And sticking out, hanging over, bending down, tangling in are hundreds of those pink-pink, white-white, fuchsia-fuchsia and variegated pink fuchsia daisy-like flowers on good stems for flower arrangements. They self sow every year so I just pluck the little "volunteers" out of the soft soil in the late Spring and replant them. Oh, I sing the praise of the cosmos! Yes, Lord, literally and figuratively.

Recently I got a book on something I know nothing about---Fibonnaci's numbers. Had never even heard of him or his numbers! Of course, I have never been into mathematics and the very word "math" has always terrified me. In 10th grade in Mt. Lebanon I had to take Geometry from Miss Nisbet. Try as I did, I couldn't "get" it. When the first report card was due, I went home and cried, "I'm going to get a 'D' in Geometry!" My mother never went to school but my angst forced her into Miss Nisbet's room. "Oh, no," said Miss Nisbet. "Sandra's getting a 'C.'" I was relieved, but I hated Math from then on. Well, this book on Fibonacci's numbers is fascinating to me these many years later. His numbers are all about the flowers and I'm into flowers! Fibonacci aka Leonardo da Pisa (c. 1170-1250 A.D.) was an Italian mathematician who developed a sequence of numbers that he found came up with amazing frequency in nature. His sequence of numbers is easy enough to be understood even by me. 1, 2, 3, 5, 8, 13, 21, 34, 55, 89, 144, 233, 377, etc. All you need to do is go: $0+1=1; 1+1=2; 1+2=3; 2+3=5; 3+5=8; 5+8=13; 8+13=21$ ad infinitum. Each consecutive number is the sum of the two numbers that preceded it. 1+2=3; 2+3=5, etc. Now why is that important or interesting to you, Sandra? Because those numbers show up with striking frequency in the arrangement of leaves, seeds and the PETALS OF FLOWERS. For instance my COSMOS. Count the pink petals, Sandy. 1, 2, 3, 4, 5, 6, 7, 8. A Fibonacci number! This white one: 1, 2, 3, 4, 5, 6, 7, 8. A Fibonacci number! All of them have EIGHT (8) petals---which, of course, is the combination number of the two Fibonacci numbers that preceded 8 (3 + 5). My lilies have two sets of 3 petals. My black-eyed Susans have 21 petals. My delphinium have 8 petals. All of them are Fibonacci numbers! Bananas have 3 sections. Apples have 5 sections. (When they get into the correlation between Fibonacci's numbers and the hallowed "golden ratio," they begin to lose me.) Even the section on the many seeds in the sunflower is fascinating. There are two intersecting spirals of seeds. They say if you count the number of seeds along one spiral, it will almost invariably be a Fibonacci number---21, 34, 55, 89 or 144. If 34 seeds curve in one direction, there will be infallibly 21 or 55 rows in the other direction!! Out of 13,000 observations of plant petals and growth patterns 96.5% of them corresponded to a Fibonacci number. That's pretty impressive, Lord. I love the way You hide Your Divine Math in everything and then wait for us, Your creation, to find it out! Hands down, one of my favorite proverbs is: "The glory of God is to conceal a thing." Proverbs 25:2 We will never find it ALL out, however. But the search. Ah, the search!!

The hundreds of cosmos have attracted pollen gatherers. They stuff their little baskets and trundle back to the hive. Often the furry bumblebee gorges himself all day on the cosmos and I find him dead the next day---stuck to the pollen.

I've got lots of gorgeous cultivated asters, zinnias, mums, marigolds still blooming, but my bees, like the Lord, prefer the least as first. They prefer the wild flower weeds to the cultivar. It's always the manger rather than the palace with the honeybee---the little, insignificant wild things as the true building blocks. It's logical. Over the millennia, the little wild ones return profusely, grow indiscriminately, ubiquitously and surely while man breeds new and breathtaking hybrids century after century. They are bred one by one, discriminately, painstakingly, time-consumingly. So they are not to be counted on. My bees have an inbred kinship with the reliable wild ones. The bees and their choices are parables.

The hive diminishes day by day. The work goes on. The remnant will remain. He always saves a Remnant. We know that from Scripture.

Occasionally when I open the hive, guard bees (c. 60) move furiously back and forth in the bottom center of the hive near the bee gate. If I close it up again for several minutes, then open it, all is quiet for a second and then they swarm to see what is going on. So I know they feel the openness and the light. They buzz back and forth for c. 5 minutes and then when they are sure there's not going to be trouble, they go about their business again. Only those bees at the bottom center show any change in behavior when I open the hive, so I know they have to be guard bees.

October 4, 1982

My poor babies. They are dwindling so quickly! Even though we're in the middle of a nice Indian summer

A Cosmos in my Kitchen

now, they continue to die. There are not many bees on the honeycomb at all.

Now they're foraging wherever they can---nectar and pollen from the cosmos still. They're even on the mums, especially the daisy mums. A few forage the asters, nasturtiums and whatever else is still around.

Haven't seen any drones in or outside of the hive for weeks. Maybe they have slaughtered them already?

Inserted Much Later: From <u>A Pocket Full Of Posies</u> by me, Sandra Sweeny Silver

"The Slaughter of the Drones." Sounds like a Machiavellian solution to worker apathy. In fact, it is just one of the many iterations in the life of the honeybee hive.

The hive of the honeybee is composed of one queen, many female workers and a scattering of males called drones. The queen lays the eggs and is, for all intents and purposes, a prisoner in her own castle. The female workers do all the work in the hive and all the foraging and gathering outside the hive.

The male drones do...well, the drones do very little They don't clean the hive or take care of developing brood or gather nectar, pollen, propolis or water. I have seen their large, furry selves obsequiously begging a female worker for a little nectar. She stops, gives him from her nectar crop and then hurries on to her busy life as he continues to meander over the honeycomb sipping here and there from the golden hexagons.

His life begins like hers in the Spring. She is born and bred to work. He is born and bred, as far as we can determine, to fertilize the queen.

One fine Spring day when the sun is blazing and the earth is pregnant with bloom, the virgin queen pokes her head outside the hive entrance. She lifts her sleek, long body into the clear air and begins to climb toward the sun. Drones from her hive and the other hives nearby smell her scent and begin to follow her. Only one drone will catch this pristine beauty. She is bigger than they and faster than most. Up, up she flies...so high the mating itself has never been observed though some old beekeepers have claimed to hear a crack at consummation.

The many drones with their powerful bodies wing toward her in a tangled comet-like horde. The fastest, most powerful catches her in mid-flight. As he impregnates her for life, part of his abdomen tears off inside her to plug his seed. He falls lifeless to the sane earth. His genes will live on for generations. His death was worth it.

The mated queen returns to the hive, begins laying eggs and never again leaves the hive unless the bees swarm. The other drones return to their hives and their rather sybaritic lives of eating, flying around and heavy buzzing until...

One fine Autumn day when the sun is waning, the nights are cold and the flowers are few, they return to the hive and the female guard bees refuse to let them enter. Too many unproductive mouths to feed over the long, cold winter ahead. Those females who tended them as larvae, fed them as adults and lived peacefully side by cramped side with them for a summer lifetime now bar them from the warm hive so that they will freeze to death in the cold, fallen nights.

On crunchy, crisp October mornings I have seen them dead in the stones and grass outside the window where my observation hive is. They haven't "slaughtered" them all. I can see a few drones inside the hive creeping close to the winter cluster. But the majority of the males are gone. A few are allowed to live. One never knows when a queen will die and a new virgin will need to be mated.

There are no parallels to be drawn to human existence, no points to be made by smug feminists. Each---the queen, the worker and the drone---has glory and gaffe.

The queen, a lonely focal point, imprisoned by her subjects and her fecundity. The worker, anonymous and servile, paradigm of sacrifice and productivity. And the drone, virile and magnificent, sine qua non, driven to a doomed consummation, slaughtered by his daughters.

October 9, 1982

I feel terrible. I just opened the hive and there on the bottom of the upper comb is a dismembered bumblebee. I missed the whole drama. He or she ventured into the hive smelling the honey because there are meager stores outside. They stung her to death (how many honeybees gave their life on that intruder?). They dismembered her for removal.

QUESTION: She's BIG compared to my little ones. How did she get through the queen excluder? The metal bars are spaced so that only the workers and drones can get up into the honeycombs. Those bars should have kept her in the bottom of the hive. My queen who is considerably slimmer than this bumblebee can't pass her abdomen through the excluder. How could this fat one get through? However it all happened, 4 or 5 of them are trying to drag the remains under the comb so they can get them outside the hive. I have been watching this task for c. 30 minutes now. They have chewed her in half and are trying to extricate the head and a wing from the comb. What amazing economy of labor! Here are thousands of bees. Yet only 4 or 5 of them have been given the task or have undertaken the task of dismemberment and removal. The rest of the bees are all totally unconcerned and doing other tasks.

At the dime store today there were 40-50 honeybees on the mums for sale. I wondered if they were from my hive. The dime store is a little over a mile from my home, so that's within their three mile foraging radius.

October 12, 1982

Was thinking that the hordes of drones that don't get to mate with the queen live the rest of their lives as virgins, monks. They can't be called "eunuchs" because they are not castrated. They are just there awaiting the birth of a new queen and the possibility that one of them may be the one to get her. I was thinking about eunuchs and their long tortured history. 4,000 years ago in the Sumerian civilization they had created eunuchs to be a sort of handmaid to the king. His testes had been cut off when he was 13 and he was no threat to the king's virility or to the queen and her maidens' virtue. If you push a balloon down on one side, it pops up bigger on the other side. Lots of eunuchs over the millennia have channeled their lost Freudian sexuality into an Adlerian will to power. Zheng He, commander of the Treasure Fleets of Emperor Zhu Di in the early 1400's A.D., is credited by some with discovering America some 50-60 years before Columbus did. He was castrated as a young boy. His immense size (7 ft.) and his huge drive for power took him at least to Africa if not America. (However, the Maya do have the Mongolian spot on the base of their spine as newborns and they do have the epicanthic eye fold of the Chinese. Who knows?) At least Zheng He (aka Cheng Ho) diligently and loyally served royalty. The Persian eunuch Mithridates was the bodyguard of Xerxes I (c. 450 B.C.) until he murdered him. About a hundred years later the eunuch Bagoas, prime minister to king Artaxerxes, murdered him in a Persian coup. Some say that the Arabs got the idea for eunuchs from the Romans. I don't believe that. From Sumer in 2,100 B.C. to Persia in 450 B.C. and all the way down to modern times with eunuchs as harem keepers. Seems pretty rooted in the Middle East to me. However, China would be a close second for famous eunuchs. Whenever it was that the Orient "discovered" eunuchs, they thought they were a pious idea. Zheng He was a reluctant eunuch. However, during the Ming Dynasty when he lived, 70,000 men castrated themselves in order to ingratiate themselves into the royal household. The emperor finally said, "That's it! Anyone who castrates himself will be killed!"

The strangest cases of castration and the most lugubrious are the European ones where young (poor) boys were castrated right before puberty to keep their beautiful soprano voices in tact. This craze and craziness seems to have appeared sometime in the 1500's A.D. in Spain. It spread to Italy and at one time the Castrati were even admitted into the Pope's personal choir at the Sistine Chapel. Now this practice was strictly forbidden under Canon Law but was nonetheless happening. Instead of calling these hapless boys "eunuchs," they were given euphemistic names like "musico." At least my drones still bear within them the possibility, if not the probability, of making music some day with a queen. They are not eunuchs or monks. They are just males hanging around hoping she will happen.

There are only 29-30 bees on the honeycomb. Precise Jesse said he just counted 33. For the first time I can fully see the stark beauty of that comb. Unlike the brood comb, the honeycomb hexagons are not really distinctly, individually capped. They are rippling, ill-defined. You can see the gold honey inside under the rippling sea of white.

I feel they are settling into their winter pattern. The hive is quiet, not humming. Most bees huddle on the brood comb. The honeycomb is deserted. On the bottom row of the honeycomb there are 29 clean hexagons that were once filled with honey. They must have eaten them. It will be interesting to see if they eat their stores this way---from the bottom up. They graduate the brood from the bottom up, too, probably because the bottom is most easily defended and far away from leaking honey.

October 15, 1982

Saw them taking nectar from the daisy mums again. Haven't seen many pollen carriers. I guess nectar is the main thing now that winter approaches. Pollen mixed with honey is mainly for brood. It's called "bee bread."

Lots of health-food-type people eat pollen. I, not a health-food type but a fascinator of nature, have a jar of it upstairs in my bathroom that I've had for several years now. You wouldn't believe the nutrients it has in it! Pollen contains up to 28% protein. Has many vitamins---A, C, D, E, K. Plus it's very high in the B-complex vitamins (riboflavin, thiamine, folic acid, niacin). When we had a ministry to kids on drugs during the late sixties and early seventies, occasionally someone would come to our house "freaking out" on acid. I would give them niacin. A doctor had told me that was a good thing to do. It usually brought them down in several hours. Anyway, bee pollen, also, contains loads of minerals---iron, copper, magnesium, etc. Every once in a while I will open my bottle of pollen, wet my index finger and stick it in the tiny multi-colored balls. I'll eat it and think, "I'm eating the seed of nature." Tastes floury, like a type of flour.

Note inserted later: I'm into sindonology. That's the study of shrouds, notably the Shroud of Turin. Blake,

A Cosmos in my Kitchen

Jesse and I went to a lecture at West Conn in Danbury on the Shroud of Turin. The blood work on the shroud was done by a hematologist in Ridgefield. The lecture given by him was extremely technical and after 2 hours, only Blake and I were still awake! He said, in essence, "These marks are blood. Not paint, but blood. I'm a Jew and I don't believe in Jesus' resurrection. I believe he was a man not a god. But I can tell you that this shroud contains someone's blood. This man bled from almost every pore of his body. His blood got on and into this cloth. How it got there we cannot now tell you. I figure in several hundred years we will know why this piece of cloth is such a mystery." It was, for Blake and me, a fascinating lecture.

The shroud has, also, been carbon-dated. But the fascinating aspect of the shroud, to me, (other than the negative image) is the pollen samples that have been found on the shroud. There have been images of some flowers discerned on the shroud, also, but those images could be the imagination of the viewers. What can't be imagined is a grain of pollen. They have taken the grains and identified the flowers from which they came. Dr. Avinoam Danin, a botanist from the Hebrew University of Jerusalem, and Dr. Uri Baruch, a specialist in pollen from the Israel Antiquities Authority, claim the pollen grains found on the Shroud of Turin "best fit the assemblage of the plant species whose... pollen grains have been identified on the Shroud (that grow) 10-20 km east and west of Jerusalem. The common blooming time of most of these species is spring=March and April." They found pollen grains from the flower Gundelia tournefortii in the forehead area of the man on the shroud. This flower blooms in Israel between March and May. (It is found in Syria and Iran, too.) The Cistus creticus plant grows only in Israel. Its pollen was, also, found on the shroud. All kinds of controversy from all disciplines riot around this shroud. Some believe the evidence. Some believe without evidence. Some scientists dispute evidence offered by other scientists. I, personally, find the study of the negative image on the shroud most intriguing. I don't need a shroud from Turin, Italy to believe that Jesus rose from the dead. But now those sweet microscopic grains of pollen my bees collect are inserting themselves into the research! Love it.

It's going to get cold tonight and tomorrow night (30's). Their foraging days are numbered.

They and all other types of bees like the cosmos. When they take all the nectar out of the flower, she begins to wilt and die. I guess nectar is part of her life support system, too.

I still have blooming: cosmos, lavender, blue salvia, celosia, asters, zinnias, marigolds, snapdragons, roses, geranium lancastriense, dahlias, impatiens, geraniums, heather, ageratum, coreopsis, phlox, black-eyed Susans, gloriosa daisies and, of course, mums. So I've tried to supply them with nectar and pollen. Also the wild clover (trifolium) is still blooming. As is the gorgeous aromatic Polianthis tuberose. I picked some late teeny wild strawberries today from the patch near the vegetable garden!

November 1, 1982

It's 75 degrees today! The bees are going crazy--- almost in a swarm mentality on the window and outside the hive. They love it! I do, too!

A lot of feces has built up on the bottom of the hive. They're busy taking it all outside the hive. They're foraging even on the dead asters. Are trying to eke out the last drams of nectar, pollen.

They are certainly dwindling in number. The comb had been full. Now looks very spotty.

The honey stores are being used up a little. There is PRECISION eating from the BOTTOM OF THE COMB UPWARDS, hexagon by hexagon. So orderly, so logical. Not a haphazard pattern of consumption. Precise.

If the honeycomb is about 21 hexagons high, then they have eaten almost half (10) of their stores as of Nov. 1. I'm not going to feed them sugar water until I see the stores are really low because I want to observe as much as possible the actual process of bees in the winter re: stores.

GENERAL OBSERVATIONS:

1. Bees are centrists:
    a. They cluster in the center of the hive.
    b. They breed starting in the center of the hive and moving sideways.
2. Bees are logical in the same way man is:
    a. They use the honey stores hexagon by adjoining hexagon in a neat, orderly pattern.
    b. They show forethought. They gather and store. They work bushes, trees, flowers intensely rather than in a dilettante manner.
    c. They work for present gratification AND future needs.
    d. They protect their power bases---the queen, the hive, the brood.
3. Bees desire much of what man does:
    a. Warmth.
    b. Shelter [protected shelter]. The choice of hive habitation coupled with their vast numbers---our "walled encampments" i.e. cities.

c. Food.
d. Meaningful work oriented toward sustenance and future generations.
e. Need for others based on survival and procreation and intimacy (?).
f. Need to procreate.
ETC.

November 19, 1982

The honeycomb is seventy-two (72) hexagons long by 21-26 hexagons high. It's difficult somehow to count the height.

As of 11/1, they had eaten only the bottom layer of stores. I HAD presumed they would eat in a totally symmetrical pattern---cell by connecting cell. For the most part, they have. But there are four (4) cells scattered here and there near the middle and top of the comb that have been eaten and are not connected at all to the main eating pattern. It's interesting how they exhibit this "centrist" tendency again in their stores consumption. A diagram would show a spike up in the middle---all empty honey hexagons.

If there are c. 23 times 72 hexagons, there are 1,656 honey cells JUST ON THIS SIDE of the glass.

November 22, 1982

It's 57 degrees. No sun. Damp because everything is still wet from the rain yesterday. But my bees are out of the hive anyway. They're all over the window, cleaning out the hive and foraging. The insiders are busy like a pre-swarm activity. Figure this out, Sandy!

December 2, 1982

It's 55 degrees and sunny. My babies are out en masse, buzzing on the window, scavenging for nectar, pollen and taking out hive debris.

December 6, 1982

This December 4, 5 & 6 have broken ALL records (records have been kept for over 120 years) for high temperatures on these days. Has been 67-71 degrees. In spite of the high temps, my bees have pretty much stayed in the hive. They probably saw the lack of nectar/pollen in the previous days.

Last Fall I dug up my dahlia tubers and have stored them in the basement to replant in the Spring. They were spectacular. Just looked at them. Now they're desiccated and pretty ugly. I remember that the ancient Maya used to eat the roots (haven't verified the edibility of such!). Plus the Aztec nobles used to wear dahlias as a symbol of royalty. A while ago, I saw The Blue Dahlia movie on late-night TV. Raymond Chandler, the great mystery writer, did the original screenplay for the film. What a great film. It definitely is a film noir (black film). I haven't seen any films in Technicolor that I would consider film noir. For me they HAVE to be filmed in black and white with low-key lighting and high contrasts between dark and light frames. Plus a film noir has to be a hard-hitting crime drama with sexy women and hard-boiled detectives like Chandler's Philip Marlowe, Mickey Spillane's Mike Hammer or Dashiell Hammett's Sam Spade. Here's a few of the black and whites I consider film noir: The Postman Always Rings Twice; Double Indemnity; Laura; most old Humphrey Bogart/Peter Lorre films and some of Hitchcock's films like Strangers on a Train.

I don't know if one of my favorites, <u>A Place In The Sun</u>, qualifies in the genre. Probably not. It would be more in the "poor boy meets rich girl, falls in love with her and kills poor girlfriend" genre. <u>The Blue Dahlia</u> and those other films take me back to a great time in my childhood. Also, I loved all the Technicolor "boy meets girl and they hate each other and then they fall in love and sing about it" films.

Was reading last night all I could about the winter in-hive habits of bees. There is practically nothing except the shape of winter cluster, temperature on inside/outside of cluster, etc. Rather banal, but necessary information. No information on day-to-day activity of bees in their winter cluster. I think that will give me impetus to observe my winter cluster more closely.

STATUS OF STORES:

If there are c. 1,656 honey cells on this side of the comb, they have consumed c. 290 cells since c. October l. That's 67 days. They have, thus, consumed c. 4.3 cells of honey a day on one side of the comb. OR a total of 25.9 (26) cells of honey a day in this six-sided hive. That seems like an incredibly modest amount of honey consumption. If I have c. 25,000 bees left, and, rounding off for ease of figuring, it takes c. 25 cells a day to maintain 25,000 bees in the winter cluster---THAT'S ONE ONE THOUSANDTH (1/1,000) OF A CELL A DAY PER HONEYBEE!!!

Assuming relatively uniform consumption on all six sides of the honeycombs, c. 1,740 cells have been consumed in 67 days (October 1 to December 6).

December 9, 1982

Now it's normal temperatures for December---25 degrees outside the hive. I've got a thermometer right above the bee gate and can tell at all times what temperature it is at the entrance to the hive---useful.

When it was warm the 4, 5, & 6th, the bees were careless on the combs. They were all over the brood comb and many at random on the stores. It's c. 67 degrees here in the kitchen.

I observe that even the bees on the upper honeycomb are "centrists." They are congregated above the middle of the cluster down on the brood comb. They all seek the middle of the hive---safest, warmest? On the brood comb the greatest amount of bees is in the middle of the comb---from the middle of the middle of the comb to the bottom middle of the comb.

The vast majority of bees in the cluster are quiescent there on the dark brood comb. Their wings are extended open, all six legs are on the comb and there is an occasional up and down movement of the pointed thorax.

The bees that are facing me on the glass (their legs and underbellies face me) those bees move more than the cluster bees. One has the impression of two to three hundred teeny legs wiggling at all times.

In the middle of the cluster slightly to the left are two fanners. Everyone gives them about half an inch (I just measured it) of space on either side laterally.

Most of the cluster is two bees deep, but there is room for it to be three bees deep. I see bees traveling up and down between the ones facing the brood comb and the ones facing me. They travel over the backs of the other bees.

Re: Honeycomb Workers

I have just been observing what these c. 62 workers are doing up there on the honeycomb. A few of them seem in charge of wax disposal. There are tiny flakes of wax that fall from the cells when they open the capped honey. These pieces of wax stick to their feet, abdomen and mouth as they pass by on the bottom of the honeycomb. They are clearly annoyed when they pick up a flake of wax and spend time shaking, pulling it off before they proceed with their life---sort of like picking up a piece of gum on your shoe and having to get it off before you can continue with your life.

So 4-5 of these bees are gathering up the flakes and doing several things with them:

1. Packing them into the propolis on the bottom of the honeycomb.
2. Nonchalantly carrying them around in their mandibles.
3. Passing them under the comb where I observe other bees from the interior of the hive come and pick up the flakes. Maybe the wax is needed in the interior of the hive or they will dispose of it.

On Testing With Their Feelers:

Twenty or so bees on the honeycomb seem to be "testing" the honey cells one by one with their feelers. Are they testing to see what the temperature of the honey is? or Which cells are ready to be eaten? or When that cell WILL be ready to be eaten? OR?? Also, some bees are cleaning out the honey hexagons just like I have observed them clean out brood cells.

A Cosmos in my Kitchen

The most activity and enthusiasm centers around the bees who are tearing open the wax on the capped honey cells. Then they sip the sweet syrup through their long, red proboscises. The enthusiasm and crowd and level of activity accompanying this reminds me of hungry dogs around a meal. As they absorb the honey, their abdomens throb back and forth, back and forth which to me indicates contentment.

I just saw an interesting thing. It is an in-hive EMBRACING DANCE! One bee entered the honeycomb and was VERY excited. She tried or actually did embrace the HEAD of every bee she met. Plus, during the embrace, she did a shaking dance. Several of the bees were too busy for the embrace and dance, but four allowed her to embrace and to dance with them. She finally found a comb with some honey in it. Inserting her head, she began to feed excitedly, her abdomen pulsating up and down, up and down. What to make of her uncharacteristic, for bees, behavior? Maybe she was telling the others: I'm here! I'm excited to get some honey! I just delivered a load of honey to the bees downstairs Or ????

No fanning in the hive at this time.

December 10, 1982

Got up this morning. It was 12 degrees outside. c. 38 degrees here in the kitchen! Get that electric heat going, Sandy! The bees were in a PERFECT "V" CLUSTER. The bottom of the "V" was right in the middle of the hive over the bee gate. The two lines of the "V" radiated out up into the brood comb.

Lovely, geometrical formation. Like a B-52 bomber formation.

I always thought it was a B52 bomber that dropped the Atom Bomb on Hiroshima. It was a B29 bomber. They were called "superfortresses." Pilot Colonel Paul Tibbets and crew left Guam on a plane called the Enola Gay for Hiroshima. They dropped the bomb, nicknamed "Little Boy," at 8:15 in the morning local time on August 8, 1945. 80,000 people were killed. They returned to Guam. The round trip flight was about 13 hours. Three days later the Allies dropped "Fat Boy" on Nagasaki killing 70,000 people. Four days later Japan surrendered. They officially signed the Surrender on September 2, 1945 aboard the USS Missouri in Tokyo Bay. My mother was against the bombing. When I was a little girl, my parents had a party celebrating the end of World War II. She was the only one at the party that felt the Atom Bombing was a bad thing. I don't know. Japan started the war. We were planning an invasion of Japan and had estimated that 150,000 of our men could be killed. We killed that many of them in order to keep that many of us from dying. Jesus said, "There will be wars and rumors of wars until the end." Most people in their heart of hearts would NEVER want a war to begin. But there are bad people out there who WANT to kill, who WANT war, who WANT to destroy. Edmund Burke said, "All that is necessary for the triumph of evil is for good men to do nothing." If William Wilberforce had not taken on the abolition of slavery in England, the Lord only knows how long it would have "triumphed." If we hadn't fought World War II (and bombed Japan), who knows what our world would look like today. AND every good fight draws blood. "Aye, that's the rub," as Shakespeare said.

December 12, 1982

We have snow now. It is 5:45 in the late afternoon. 20 degrees outside. 63 degrees here in kitchen.

A Cosmos in my Kitchen

They are in a cluster, but not the tight, perfect "V" which I saw the other morning. They are clustered in the middle of the hive from the top to the bottom of the brood comb in an eccentric "O"-shaped cluster.

NO bees on the honeycomb. All are in the cluster. Very little movement. Fifty percent of them are totally quiet. Fifty percent with minimal movement of the legs---no traveling.

Actually, the winter cluster reminds me of the swarm cluster. It, too, is compact, tight, organized, minimal movement. "All for one. One for all." My sweet little ones.

A part of the thorax of a drone, no doubt stung or starved to death, remains on the lower ledge of the honeycomb. He's been there for months. Too big a chunk to extract through the space, so he remains---wedged, mute reminder of all who are "indispensable."

Haven't seen the queen since late August. They must have her on an inner comb, safe from possible harm.

Since no bee is on the upper comb, I just got a relatively accurate count of the empty honey cells. About 310 now. So times the 6 sides. About 1,860 empty cells. For 73 days since October 1, that's an average of 4.24 cells of honey consumed a day per side. My other average was 4.3 cells a day (See 12/6 entry.). So I think I'm relatively accurate in this estimate of rate of consumption of stores per day. Six sides times 4.24 equals 25.44 cells per day for the whole hive.

If there are 78 days left until March 1 at 25 cells per day consumed, then 1,950 more cells will be consumed by March 1, 1983. These calculations assume uniform consumption which may not be the case, plus they do not take into consideration dwindling.

My previous calculations showed c. 1,656 total cells per side times 6 sides. That's 9,936 potential cells of honey in the hive. They need 1,950 more cells of honey which will bring us to March 1 and potential foraging weather.

From October 1 to March 1 (161 days) at 25 cells consumed a day, that's 4,025 cells of honey stores needed for that five month period in order for them to survive well.

SO if there are c. 9,936 cells of capped honey to take them through a five month period AND they actually need only 4,025 cells to survive well, THEN they lay in over TWICE THE AMOUNT OF HONEY THEY NEED TO SURVIVE THROUGH THE WINTER. "Consider the (BEE), thou sluggard." Proverbs 6:6 God's Word.

I assume they will consume more the colder it gets. Plus, they will need honey for the developing brood and their increased work load when the queen begins to lay soon.

December 15, 1982

6:00 P.M. I haven't examined them for three days and in the lower right hand corner of the hive on the bottom of the hive is standing water!! It stands 3 and a half inches long and a quarter of an inch high! Now WHERE did that come from? It has been zero at night and c. 20 degrees in the daytime for several days. Yestereve and today the temperature thawed to 32 degrees at night and 45 degrees today. So maybe it was something in the hive that froze and thawed? It's right there where the old queen cage is still stuck. Will have to watch this. Hope it's not a harbinger of trouble. The bees aren't paying any attention to it and maybe they need some water to drink.

A Cosmos in my Kitchen

Just watched another of those EMBRACING DANCES on the honeycomb. She enters excitedly. She moves quickly. Goes from bee to bee. She has embraced and shook 23 of them so far. Have to stop and get Jesse's dinner.

I've been reading A.A. Milne's books, <u>Winnie-the-Pooh</u> and <u>The House at Pooh Corner</u> to Jesse. Now there was a honey-loving little bear! His songs in praise of honey are so cute.

"Isn't it funny
How a bear likes honey.
Buzz! Buzz! Buzz!
I wonder why he does?"

Pooh whose full name is Edward "Winnie the Pooh" Bear is obsessed with finding and eating and storing and getting honey. What a fun time Milne must have had writing those classic books! What a fun time we adults and children have reading them!

11:30 P.M. I'm back. The Embracing Dance. The Embracing Bee clasps the Embracee with her forelegs on their heads, abdomens, wings, stomachs, sides, anywhere she can clasp. She shakes them.

The bee embraces no special place on the other bee's body. When she has "got" a sister, she shakes her.

Some of the bees' reactions:

1. Some put themselves in her path to be embraced and shook.
2. Some were just there and allowed the embrace and shaking.
3. Some seemed to be ill at ease with it, but tolerated her.

4. Two of them tried to get away. One succeeded. The other didn't and was embraced and shaken.

Could she be a groomer of others? The embrace is so brief---2-3 seconds---and so random, I don't think so. She, however, GROOMS herself after each embrace. She grooms her antennae, forelegs, proboscis (this is not OUT during the embrace). Another in-hive "dance." (?)

Blake made an observation that may be valid. He said, "It's easy to tell which honey cells they'll harvest next. They are the ones where you can SEE the honey under the wax." He may be right because the ones that are half-consumed (have been tapped) are all white white, see-through. I'm going to watch this and see if his observation bears out.

I love it when Kathy, Blake and Jesse make observations, too!

December 22, 1982

Always when I open the hive, I feel the glass and it is consistently warm:

1. At the bottom near the bee gate it is warmest.
2. In the middle of the hive it is warm, but not as warm as the bottom.

I conclude that they heat the hive from the bottom up. They, also, stop the cold right where it comes in at the bee gate.

It is 32 degrees. 70 degrees in the kitchen.

About fifty (50) bees are on the honeycomb. The cluster on the brood comb is loose. They are warm, busy. There is a lot of activity in the CENTER of the cluster. They seem to be feeling their cheerios because the sun is out and they are warm.

Re: The 12/15 entry on water in the hive. The water has now receded from the floor of the hive and is currently banked up against the queen cage. I still haven't figured out where the water came from, why it is there.

There are certain bees who clean the cells on the upper honeycomb. That's a job because there is a lot of debris once the cell is void of honey. There are the many bits of wax from the covering and the residue from the cell once it has been cleaned to use again. So the honey cell cleaners are busy.

December 27, 1982

For Christmas Steve got me a magnifying glass with a light on it. Now I can see minute details much more clearly. For instance:

I can see that NONE of the bees on the honeycomb have damaged wings. The bees who have been outside foraging have tattered and torn wings. But these wings are pristine, clean. I deduce that they have done little or no foraging, were born in the autumn and didn't have to forage, have been mainly hive-bound.

After all these months, the drone's carcass, though dismembered, is still stuck on the bottom of the upper comb. They constantly pick away at it and shift it back and forth. Often they will make an attempt to drag it under the comb, but it is too big to fit under that small space and they give up. They are preoccupied with trying to remove his body.

January 4, 1983

1983! A new year with my honeys!

It's 39 degrees now, but it drops into the teens at night and in the early morning. It is Noon now.

There is MUCH condensation in the hive: on the window and all over the floor now! In fact, fifteen (15) bees are trapped, in a sense, on the bottom of the hive. They're soaked and can't get up the glass no matter how hard they try and they do try. This is a serious problem for them. I think I'll open the false door on top of the hive and leave on an observation light that I have installed on top of the hive. Maybe these measures will help to release the condensation. I hope so.

2:50 in the afternoon. Most of the condensation is gone from the glass. About fifty (50) bees are immobile, drenched and appear to be dying on the bottom of the hive. Their bodies have absorbed a lot of water. Their abdomens are pulsing, so I know they are alive. But they have a wet brown color and are all in a huddle in the bottom center of the hive where the greatest concentration of water was. The other bees are going about their business as usual as these babies are dying. Are these workers guard bees or "sacrificial" bees whose duty it is to absorb the flow of water in the hive to protect the others? Or are these just bees who have managed to become enmeshed in the water?

They blocked the side vents with propolis last summer, so I can't even put a fan near them to try to dry out the hive.

A Cosmos in my Kitchen

I feel so helpless! Needless to say, there are about thirty (30) bees fanning in the hive like crazy. We need fanners!

Fanners will fan to maintain the temperature of the hive as well as to fan off water from the nectar and water in the hive. Fan, little ones!

6:00 P.M. The pile of bees, soaked and apparently dying, grows.

January 6, 1983

The Lord is great! He is so good! His mercy endures forever and His power is to the grave and beyond!

Today Lena, my au pair from Sweden, Jesse and I witnessed one of the most amazing miracles I've seen in my 28 years as a Christian!

I have been very prayerful about the condensation in the hive that was destroying bees. Today many, c. 150, were dead in the water at the bottom of the hive. Water beads were all over the glass and I could see water on the brood comb even. The bees were very sluggish. Many were falling and becoming mired at the bottom of the hive. Even bees on the honeycomb were dead and dying. My hive was dying right before my eyes.

I called Ed and told him about the situation. He couldn't believe the amount of water. After a long discussion in which I impressed on him the FACT that my bees were indeed dying, he suggested getting silica-gel, scraping away the propolis from the vents and putting the desiccant there to help absorb the water.

In the meantime, I put a lamp near the hive. I was hoping the heat would help dry up the water. A lot of the water left the glass, but the bees continued to fall dead on the comb and drop to the bottom of the hive. All of us were praying for them. I went to write a letter to Keith and Carole. Two hours later, I checked the hive and ALL my bees, except 40-50 were inert, dead in a giant, pyramidal heap on the bottom of the hive!

Thousands and thousands of bees with NO MOVEMENT, not even their little pulsating abdomens or wiggling legs or feeble movement of feelers. DEAD! At first I thought maybe I had killed the bees with the light, but I felt the glass and it was barely warm. Besides, the heat was not on in the kitchen and it was 40 degrees outside, 50 degrees inside.

I searched with the magnifying glass for a fragment of life among all those bees. Lena kept asking, "Anyting? Anyting?" Nothing. All were in contorted, rigor mortis-like shapes. Heap upon heap of dead bees. Lena searched. There was no question they were dead.

I called Ed, but he wasn't home. I talked to his wife, Nita. She was flabbergasted, too. What could have killed all my bees so suddenly? I knew it HAD to be the condensation, the excessive moisture. They were in a steam room atmosphere.

I put my hands on the glass and Lena and I prayed, "Oh, Jesus, You Who raised Lazarus from the dead and Who came back Yourself, please bring my bees back to life." I implored Him.

Then Lena and I went to the store and picked up Jesse at school.

We were gone about an hour and a half. When we came home, we immediately looked at the bees. Even those who were alive before we had left, were now dead in a PILE on top of the giant heap. I saw about 12 bees with some life in them on the comb, but it was clear they were dying, too. I had to find out if the rest of the hive was dying. Maybe this was a phenomenon happening only on the observation side of the hive.

I went outside and climbed up on the high stool that I had left positioned there when I had re-hived the clipped queen after aborted swarmings. The bee gate was clogged with dead bees! A huge mass all stuck together. I knew it was not the lamp. It was the water. Those bees were orangish and very black like the other bees who had absorbed a lot of water. I had Jesse hand me a stick. I scraped the dead bees away from the gate. A GUSH of water ran out onto the window sill! Their dead bodies had clogged the entrance. No air was getting into the hive. PLUS their mass allowed NO water to run out of the hive.

I got a flat board and pulled out most of the dead bees I could reach. I scooped out all the water I could. I went back into the house and scraped away some of the propolis from the side vents. I poked holes with a needle in the rest of the propolis, as Nita had suggested.

There were still 8-10 bees on this side who showed some life and I thought maybe the queen was still alive somewhere in the back of the hive. There was a remnant that could be saved. I prayed again.

Now the following sounds hysterical and incredible and it is incredible and miraculous, but... All the preceding happened from 2:00 in the afternoon when most of the bees were dead until 4:00 in the afternoon when I unclogged the dead bees.

By 4:15 THE BEES HAD COME TO LIFE AGAIN! I mean it! They were literally resurrected from the dead! One by one, they re-animated. That which was dead was alive. Jesse, Lena and I witnessed this miracle. After about 20 minutes of awe-struck viewing of mass resurrection, Lena ran to the carriage house and came back with her camera. "They never believe, but we get pictures."

Steve walked in the back door as the three of us were rejoicing and yelling. We told him. He half-believes us, is hungry, tired. So goes the world of miracles!

It is 6:30 now. I praise You, Lord! All but c. 60-70 bees at the bottom of the hive are not only ALIVE, but they are in swarm fever!! The hum can be heard all over the kitchen. (Are they praising You, too?!) They are hyper-active. Every fiber of each of their little beings is moving. Such is the resurrected life!!

I, Sandra Sweeny Silver, witnessed this mass resurrection from the dead on (my sister Tooie's birthday) January 6, 1983 from 4:15-6:30 P.M. And it is still going on.

8:20 P.M. Now for the "scientific" observations on the re-born hive, though all the preceding was "scientifically, empirically" observed. We SAW. I FELT the dead bees with my own hands. I HEARD no sound except the awful silence of death. So, minus taste and smell, all that we observed empirically happened.

They are still in a frenzy of activity. Active ones are at the bottom of the hive trying to raise to life through pushing and stroking those at the bottom of the hive who are still dead---even to Steve's eyes.

Through the open vents, I just smelled the inside of the hive---that wonderful exotic, cinnamon and spice smell. And the air inside smells cool. That wonderful smell is like cinnamon gum, appetizing. One would imagine it would smell like honey, but it doesn't. The glass is warmer than usual from all the activity. I would think they were swarming were it not January!

One bee just picked up a dead bee by her thorax and carried her 2 inches to the bottom right of the hive where there are a pile of "dead" bees. After her inert body was deposited, I saw her legs begin to wiggle.

The water is cleared away from the middle of the hive all the way to the right corner. Standing water remains in the left of the hive about 5 inches in toward the middle. There are two bees at the extreme left of the hive fanning. One is on the wet floor fanning. The other is upside down on the brood comb, wings to the hive floor. Keep fanning.

Possibility: They were in a state of suspended wet animation? How would that work? Can insects go into or be induced into a state of suspended animation?

1:30 in the morning They are still busy and active. The hum has moderated a little, but everyone is still happy.

Interestingly enough, they are stripping away the remaining propolis from the vents with their mandibles. I started the process with careful knife scraping and a needle. They are finishing it. They have been working all evening on this task.

There are only c. 30 bees left "dead" on the bottom of the hive, but they keep working on them, stroking them, pushing them. I think the stroking is an attempt to absorb the water from their bodies. One by one, as He resurrects us, they are being resurrected.

I know most will not believe that the bees were actually dead, but I KNOW they were and I am grateful for and awe-struck by this miraculous demonstration of Your power. Amen. (Unless they were in a state of mass suspended animation?)

January 7, 1983

The day after the miracle. They are healthy, active. I see they have moved some pollen stores that I saw on the brood comb when they all lay dead. They apparently don't feel they are safe here or they are needed in the back of the hive.

At the bee gate where I scooped away hundreds of dead bees yesterday there are many more there today. Alive, I presume.

They have eaten away all the propolis from the screen vents. There are still dead bees at the bottom of the hive.

Thank You, Lord, for all this activity.

Note inserted later: I've researched a lot and can find NOTHING on animals or insects going into a state of suspended animation. Suspended animation is "a state of temporary cessation of the vital functions." Bears hibernate, of course, and frogs can live under the ice in a pond for months. Their vital functions are slowed down. They do not cease. Fakirs in India have for centuries been able to regulate their heartbeats and digestive systems. Some of them have had themselves buried in the earth for months. Unfortunately, only a few of those fakirs survived. But the several that did survive have proven that the mind can control the body's vital functions. Technically that's still not suspended animation. In suspended animation there is no heartbeat, no breath. It is temporary death. My bees were dead. And then they were so alive they were rioting in the hive. It was either suspended animation or resurrection.

In "Star Trek" episodes and in <u>2001: A Space Odyssey</u> the writers put star-travelers in suspended animation so that they didn't age over the eons of time it took them to reach outer space. Man has been on a perpetual search for PHYSICAL IMMORTALITY since he forfeited and was expelled from Eden. Suspended animation does not qualify in this Fountain Of Youth category, but cryonics (all in the news now) does. Cryonics (coming from the Greek word "kyros" that means "icy cold, freezing") is the practice of freezing clinically dead people until a cure for whatever killed them can be found in the FUTURE. The scientists who do this postulate a shaky hypothesis: I die of cancer. I pay you big bucks to freeze my body and MAINTAIN it for decades or hundreds of years until a cure for cancer is found. Then they find my body in that capsule and what? Do they unfreeze my body and pump me full of whatever is the cure for my cancer? How does that work? And when I was frozen, Dr. So-And-So knew me and did the procedure on me, but, unfortunately, Dr. So-And-So died 93 years ago. So for those 93 years many generations of doctors have vigilantly guarded my body just for this time? Does anyone in his right mind actually believe this nonsense? Now, there may be something to freezing bodies. But signing up to be frozen and assuming your stiff body will be taken care of in perpetuity is crazy. Case in point: Plutonium has a half-life of about 24,000 years. We're building all kinds of nuclear reactors. Who's going to watch plutonium residue for 24,000 years after we're through with the nuclear reactor? I was in Mt. Lebanon, Pa. when they built the first nuclear reactor in the United States in Shippington, Pa. in 1957. Did scientists even contemplate the half-life of plutonium at that time?

Of course not. They told us we would pay pennies for the electricity it would generate. Did they even contemplate who would be WATCHING the deadly radiation from that plant for 24,000 years? Of course not. Well, who is going to watch my cryonically frozen body for even a couple hundred years? I'm not into the SCIENCE of it all. I'm into the PRAGMATICS!

Note inserted much later: Listen to this from the Cryonics Institute in Clinton Township, Michigan. "...a member patient is infused with a substance to prevent ice formation, cooled to a temperature where physical decay essentially stops, and is then MAINTAINED INDEFINITELY (my capital letters) in cryostasis....At an OPTIONAL EXTRA (AT EXTRA COST) a Cryonics Institute Member may contract for cryonics professionals from Suspended Animation, Inc. to wait by the bedside during a terminal condition and immediately begin cooling and cardiopulmonary support upon pronouncement of death....When AND IF future medical technology allows, our member patients hope to be healed, rejuvenated, revived and awakened to a greatly extended life in YOUTHFUL good health, FREE FROM DISEASE OR THE AGING PROCESS." In 2007 they have 77 people in "cryostasis" which means they have 77 frozen people in capsules that they are looking after FOREVER. I think their phrase "maintained indefinitely" pretty much means "forever." (I notice they do have a Potluck Dinner coming up soon.) If I would choose their Option One, it costs $28,000. Option Two is $35,000. But if Suspended Animation, Inc. gets involved the Options go way up to $88,000. (#1) and $95,000. (#2) I can't imagine having this procedure done WITHOUT those suspended animators. They are at the bedside when I die. They begin the cryonic process immediately. They transport my body to Michigan. I'd spring for the extra $60,000. Without them, I'd decay on the plane to Michigan where my capsule awaits! Sudden death while traveling in

A Cosmos in my Kitchen

Croatia? They have a London connection. Want to have your dog with you in Michigan? They already have 48 pets in cryostasis. Their sales pitch ends with "But please—don't wait too long. THAT CAN BE FATAL, AND OFTEN HAS BEEN….Help us to share and build THE LONG TOMORROW." (My capital letters.) Suspend the interesting science of it all. I'm still into the nitty gritty pragmatics of WATCHING the bodies. Some cryonically frozen people HAVE ALREADY decayed because of electrical/etc. problems inside the "capsule". They were removed and buried. (Refunds?) Whose going to be around to make sure the building and my capsule are not accidentally destroyed by, say, urban renewal in the next several hundred years? AND the "Long Tomorrow" is only guaranteed IF you have generation after generation of scientists committed to watching and regulating and protecting your expensive capsule. What kind of money would the Institute have to PAY FORWARD in order to get that level of commitment and vigilance?! To me, it's pure insanity. To put it kindly, the "encapsuled" were when they were "not in capsules" short-sighted and misguided people. The logic and the PRAGMATICS just aren't there.

January 8, 1983

In spite of:

1. The fact that I scraped away dead bees from the gate and released water from the hive.
2. And the 36 hours it took them to clear away all the propolis from the four air vents (2 on the sides of the brood comb; 2 on the sides of the honeycomb) and thus create good, clean circulation.

In spite of the good circulation of air and the unclogged bee gate, THERE IS STILL EXCESSIVE MOISTURE IN THE HIVE.

There is a lot of condensation that runs down on the glass and there is standing water on the bottom of the hive.

After the resurrection, there were only 40-50 bees still dead on the floor of the hive. Now that number has grown to several hundred. I think this MAY be normal dwindling, but I am not sure given the water problem with this type of indoor hive.

Today it is in the 40's, sunny. I took Jesse's toy binoculars upstairs to our bedroom and looked from the 2nd floor bay window into the bee gate on the 1st floor. The bees were taking out dead bees and dropping them right outside the hive on the window sill. Some were removing other debris. What a massive job it's going to be to lug all those water-soaked bees out! They already have to carry their own weight in order to take one bee out let alone take one bee plus all the moisture that bee has absorbed. In the hive they are piling some of the dead bees in the right hand corner next to the queen cage.

The remaining bees are back to winter normal after those two days of resurrection frenzy. The winter cluster on the brood comb is half the size it was several weeks ago.

When I smell through the air vent, it is still the cool, cinnamon, spicy smell. Nothing I've ever smelled is exactly like it. Now that the vents are clear, that aroma permeates the kitchen.

Thank God the hive is normal again. I suspect my queen (whom I haven't seen since August) is going to start laying soon for the spring brood.

They are not touching the sugar water I gave them. Saw 4 or 5 small bubbles of taste when I first put it in, but that's been all. I can occasionally see their little nectar tongues go up into the hole on the nipple. Then a tiny bubble will rise from the bottom of the bottle and float slowly up to the top. That's a sign they're feeding.

The constant search is for higher magnification to observe my bees. I just got a new magnifying glass with a 5 power bi-focal on the right hand side. On my other one the light wouldn't work. This one has a flashlight bulb and will switch on and off. I LOVE it so far!

January 12, 1983

I'm coming to the last page of my First Bee Book. This journal has been filled with building, breaking down, life, death and resurrection. In this, just one facet of God's infinite number of creations, is the microcosm of the whole of His Work.

I am awed at the complexity AND the simplicity of the bee's life and see her existence as somewhat analogous to our own.

The dwindling continues. I see them preparing the cells for new life. This year's first fruits have to be remarkable coming from and tended by born-again bees.

Selah.

January 13, 1983

The last book was a regular-sized bound book with a maiden and a unicorn on the front---from the Unicorn Tapestries in the Cloister. The myth is that only a virgin can tame a unicorn, that ultimate phallic symbol. I have no wish to tame my wild ones and having had two husbands and three children I have little chance!

This new observation book, My Bee Book Two, Too, is half the size of the other. It is like the little diary books I had as a child. Wish I still had just one of them. It's cloth bound, navy blue with white flowers and red leaves. Feels nice. Wonder what wonders it will spell?

On this first page I've stuck a bit of propolis from the vents. I chewed it for a while. It was wonderful. The smell from the hive is cinnamon and the taste of this gummy substance is like Dentyne gum, hot and cinnamony. So the wonderful spicy smell that wafts through the kitchen is the smell of this amazing substance. Will have to read up on propolis and see what I can find out about it.

I've seen some chewing gum with propolis in it. I guess the bees mix that sticky resin with some of their wax in order to get this chewiness. That's logical because resin is so sticky and the wax blended in would make it more malleable. Bees have to do so much with the propolis that they need it to be a workable substance. The lining of the brood cells with the propolis has to create a very antiseptic environment for the brood.

Again my honeys are defying the books and all dicta. It is sunny, minus 2 degrees F. with the wind chill factor, but 24 degrees on the thermometer above the hive. There is no sun directly on the hive---gets morning sun. It's 12:30. BUT they are out of the hive

A Cosmos in my Kitchen

en masse---flying around the entrance, buzzing to beat the band and congregating at the bee gate! I've got to buy a pair of real binoculars to check this out from the upstairs window. I've been using Jesse's play one for months and it just isn't good enough for me to see the bee's knees. Again this batch of bees shows that anything is POSSIBLE with apis mellifera. They are not supposed to be able even to flap their wings in these temperatures!

The bee has an amazing and efficient flight "machine." Not only do her four wings propel, they, also, steer. When she flaps her wings, they can stroke up to 11,400 times per minute! This intense flapping is what produces the bee's "hum." She can fly forward, backward, up, down or hover in mid-air. Men observed the flight of birds, longed for such freedom and invented the airplane. Men observed the versatility of wing movement in the insect world and invented the helicopter.

I've always loved the Greek myth of Icarus and Daedalus. Daedalus was totally brilliant. He took the spine of a fish and fashioned it to a piece of metal and invented the saw. It was he who designed the legendary labyrinth to imprison the Minotaur (a half-man/half bull monster) in the bowels of the Palace at Knossos in Crete. How perverse was the Cretan king's wife Pasiphae! She had Daedalus make her a wooden cow to hide in so she could have intercourse with a bull. She gave birth to the Minotaur. It had to be hidden. Voila, the labyrinth. Then there's the Theseus myth that intertwines with Daedalus' story. Helping Theseus is what got Daedalus and his son Icarus imprisoned. While awaiting his doom, Daedalus would watch the birds from his prison window. Being the soul of ingenuity, he said, "King Minos may be king of the land and the sea, but he is not king of the air. I can escape from prison and death if I can take to the air." Daedalus fashioned wings for himself and for his son Icarus. He gathered feathers, made an infrastructure and then affixed the feathers to the "wings" with beeswax. Daedalus told his young son, "When we are flying, Icarus, don't soar too close to the sun or the wax will melt and you will fall and be killed!" But do the young always listen to their parents? Icarus was exhilarated by the joy of escape and the freedom of flight. Up and up he soared toward the sun. Of course, the beeswax melted and the feathers came off and Icarus fell into the sea and drowned. It's mind-boggling to me that a part of the Mediterranean off the coast of Crete is STILL CALLED THE ICAREAN SEA!! If Daedalus was one of the first men to attempt flight, that was in c. 1450 B.C. That's c. 3,500 years ago. I've always believed that in all myths there is a kernel of truth. Probably there was an artificer in antiquity who did build for himself and his son artificial wings. And the fact that the sea is still called by the dead son's name shows me that his son did die in that first recorded attempt at flight. People like me who try to figure out what the real historical meaning (kernel of truth) behind a myth is are ones who believe myths are not just made-up stories, but stories that have their bases in a reality. (The etiology, for instance, of Ulysses' encounter with the giant Cyclops Polyphemus, and the eye-piercing and the ensuing eruption could be a description of an eruption of Mt. Etna.) I can't keep writing this stuff down. Have got to start dinner and then play tickle bug with Jesse.

6:00 evening. This cold weather has taken most of the condensation off the glass, thank goodness. They have consumed only about one tenth of a bottle of sugar water since I put it in last week.

The flurry of activity at 12:30 lasted about 15 minutes. I wish I could have SEEN what was going on. I went to Caldor's and bought a pair of binoculars, but they didn't give me a close enough view of the bee gate.

I'm used to the resolution of my magnifying glass and I guess that has spoiled me. But I'm going to keep looking for a spy glass or something that gives me magnified viewing from the outside. I can look into the bee gate from upstairs and, also, from the bay window in the Victorian living room.

They are very busy, excited now that I've had them open for a while. So I'll close up.

Still the smell inside the hive is delicious and now that the cold weather comes through the bee gate, the smell is cool, minty.

I thought bees were supposed to die if they were out in 24 degree weather. My bees sure are unusual because five or six thousand of them were out there today!

January 16, 1983

Well, Steve has flown to Dallas. Blake and Lena have gone skiing in Vermont. Jesse is at Robbie's for the afternoon. Kathy is away at college. I am alone and can do some of the things I enjoy. One is watching my honeybees.

I do want to say something else about Daedalus, though. He's one of my favorite mythological characters. I believe there was a genius inventor like him who was Greek, who served King Minos on Crete and who attempted flight with his son. Daedalus is, also, credited with solving the riddle of how you can thread through a spiral nautilus seashell. Many others had attempted to make a thread that was strong enough to go through the shell and out the other end. Daedalus, being the genius he was, just attached a thread to an ant, put the ant in the shell and the ant found its way out the other end, dragging the thread behind it. That's the type of genius that thinks outside the box!

Alexander the Great was not the mental genius that Daedalus was, but when he encountered the Gordian knot, he showed pragmatic genius. He, of course, went on to conquer the world all the way to India. The phrase "Gordian Knot" is still used in academic circles to describe a thorny question or situation. The actual Gordian knot was in Phrygia (Turkey). It was a massive knot made of rope or bark. Many, many over the centuries had tried to unravel it. It was said that he who was able to untie the Gordian knot would rule the world. Alexander the Great in c. 333 B.C. was wintering at Gordia. Of course, they wanted the young king to attempt the feat. Alexander examined the knot, tried a few times to unbind it and then picked up his sword and cut it in two. "There," he said. "I untied it!" Now that's a type of genius. A "Gordian Solution" is an ingenious solution to what has been a continuing, perplexing problem. Of course, Alexander when given the choice by the Delphic oracle of "A long life filled with happiness" or "A short life filled with glory" picked the latter. Dead at 33. I'm still writing in this journal about him. That's a type of glory.

The bottom of the hive was uniformly covered with dead bees (normal dwindling, I suppose). In the last 24 hours, they have removed or put to one side every bee that was in the bottom center of the floor. It is now clear for a good 5" swatch. That must mean the entrance and egress to the hive is right there and they want to keep it clear. With my magnifying glass with the light on it I can look through the whole bottom of the hive and have confirmed that it is, indeed, totally clear all the way through the three combs. I see that there are dead bees in the whole hive on either side of this clear path, however.

A Cosmos in my Kitchen

I'm now going to count the number of empty honey cells and see how it compares with the last count.

On this observation side there are c. 455 empty cells now. Last count, 12/12/82, there were c. 310 empty cells. Ed told me there would be MORE empty cells on my viewing side than on the other unseen sides because the bees eat from warmer to cold. But for my rough, unscientific calculation, I'm going to assume uniform eating pattern on all 6 sides. I'll bow to Ed and say 400 empty cells on all sides rather than 455. Six sides times 400 empty cells equals 2,400 eaten cells.

12/12 to 1/16 equals 35 days.

Last count there were 310 cells consumed this side.

This count 450 cells consumed this side.

450 minus 310 equals 140 additional cells consumed this side since 12/12.

35 days divided into 140 cells this side equals 4 cells per day consumed on this side. That's right in the range of the other calculations!!

This is exciting. Still appx. 4.0 cells used on each side per day or c. 24 cells of honey consumed per day in the 6-sided hive!

1. 12/6/82
    4.3 cells per day per side
    25.9 cells per day whole hive
2. 12/12/82
    4.24 cells per day per side
    25.44 cells per day whole hive

3. 1/16/83
   4.0 cells per day per side
   24.0 cells per day whole hive

NOW--there have to be some corrections.

1. I don't have as many bees as I had in the first two calculations due to continued dwindling.
2. However, I believe the rate of consumption has remained uniform because: the queen may be laying. The honey the dead bees would have used is going to developing brood and for strength to tend the brood.

BUT it is amazing to me that for whatever reason, the RATE of honey consumption has remained the same even though the status of the hive has changed.

We've had the first snowfall of the season---6 inches. It's beautiful and LONG awaited. Outside the hive the temperature is 29 degrees.

The sugar water I provided for them goes down by millimeters. They sip occasionally. They obviously prefer the sweet, liquid gold. Who wouldn't?

The condensation problem on the window is over. It's clear. However, there is a lot of water on the outside of the wax capping on the honeycombs.

Blake's observation, made months ago, is right, I'm sure. He opined that when the honey is just ripe for hive consumption, the wax turns from clear to chalk-white. They then eat ONLY those cells with the chalk-white coloring no matter where they are found on the comb.

THERE IS NOT ONE CELL WHICH HAS BEEN CONSUMED OR IS CURRENTLY TAPPED THAT DIDN'T HAVE OR DOESN'T HAVE THE CHALK-WHITE COLOR.

I imagine it is a function of when the cell was filled with nectar. Those filled in July and August will ripen and turn chalk-white before those filled in Sept./October. They, also, wait until the WHOLE hexagon turns chalk-white before piercing the wax. Some are partially that color. They will soon be ready.

This is a visual thing for the human observer. I suspect it is VISUAL and TACTILE for the bee observer. The chalk-white ones must feel crustier to their feet, antennae. I haven't read about this. Maybe Blake's is a new observation?

They have worked very hard to make the brood cells clear, smooth, free of debris for the new batch of bees. With my bifocal glass and the light, I see they are beautifully polished mahogany houses, shining from the inside out.

Unlike the honey cells that have debris and bits of wax stuck and strewn all over the insides of the cells, the brood cells are spic and span. But I'm sure when the nectar begins to flow again, these honey cells will be sparkling.

Checking the screen vents, I noticed they pushed out a sizable piece of good propolis. I'm chewing the good bee gum as I write. Yum.

Now I've taken the propolis out of my mouth. It's waxy, gummy. I'm mashing it up. It's stickier than some gum in that it sticks to your teeth. Now that I've chewed it, some of the dark reddish brown color is gone. I'm putting it back in my mouth. LOVE that Dentyne gum taste and the taste lasts longer than

Dentyne's does. It does stick to my teeth, though. I look at it. The dark color is lighter and lighter as I chew and suck. I'm putting it right here on the page under CHEWED SPECIMEN OF PROPOLIS.

January 18, 1983

It's the coldest night of the winter---minus 5 degrees. It is cold here in the kitchen. One of my electric heaters died.

Carol F. told me yesterday that she wanted to get some bees, too. But she wants an outdoor hive. I told her to call Ed. She did and ordered a package for April. Carol is a good artist. She was showing me some of her paintings the other day.

I am not really evangelical about bees as I am about the Lord. I don't believe they are for everyone as is He. However, I am verbal and enthusiastic about them. Who will be caught will be caught.

Was looking at some of the paintings in one of my many books on art and saw Peter Bruegel's "Landscape with the Fall of Icarus." Just wrote about Icarus. The painting is mainly landscape. But way off in the distant sea, you see a splash where Icarus is landing. Love the Bruegels!

This afternoon c. 3:00 the bees were making a "busy" hum. I opened the hive and they were doing a kind of Group Dance. The dance was as follows: Hundreds were on this side all facing upward to the honeycomb. They were grouped in loose vertical lines. They moved rapidly up and down, up and down, up and down very excitedly. I have not observed this particular Group Dance before.

I didn't have time to really examine what was going on. Perhaps this was a response to a message from their queen OR to the cold weather OR, akin to the weather. Perhaps it was a collective exercise to keep the old body warm and limber Or to generate some quick burst of heat in the hive Or...

Right now, 9:00 P.M., there are c. 300 on this side. Looks very bare with only that many. They are not in a tight cluster, but are close together. The cluster looks like a submarine, horizontal with a dorsal fin. There are no bees on the honeycomb.

They have consumed almost all of the sugar water. That's three quarters of a bottle in 12 days---very modest.

I got some comb honey yesterday. It's much darker in color than the golden combs of my hive bees, but it's very good and strong-tasting. I prefer the taste of honey in the comb. I scoop a nice spoonful, put it in my mouth, suck on the liquid gold and then chew the wax. Good.

Everyone's getting so health-conscious. Hate that concentration ONLY on the body. Yes, the body and health are important. Vitally so. But give as much care to the spirit and the mind as you do to the body! Body buried, decomposes, turns to dust. Spirit and mind go on forever! BUT if you want a cholesterol free, salt free, fat free food, eat honey!! You could LIVE on honey. Seriously. It has everything the body needs to sustain life including water!

John the Baptist lived on locusts (protein) and wild honey and look where it got him! Jesus testified that "among those born of women there has not risen anyone greater than John the Baptist." But He went on to offer hope to us spiritual midgets: "Yet he who is least in the kingdom of heaven is greater than he." (Mt. 11:11)

January 20, 1983

They are doing that GROUP DANCE again! This time I'm observing it more closely: 5:30 P.M.

1. Their wings are fanning as if in flight.
2. They are facing me on the glass. All six of their legs are moving up and down on the glass but they remain in place.
3. They are all pointed upward and the Dance becomes so frantic that many fall to the bottom of the hive and then have to climb back up again. As they climb, they continue agitating and dancing.
4. I believe this could be just a mass exercise class to keep the wings and legs in good working condition for the work ahead. They all participate and seem to "enjoy" it.
5. It is not related to the hive being open as the noise of such activity attracts me to the hive and then I open it.
6. It is accompanied by a humming but not a busy, contented hum such as in the summer. The hum seems associated with hardship, exercise---sort of like bee "grunts."

February 2, 1983

My. They have been humming and busy for these last two weeks. I take this to be evidence of egg laying.

The minute I open the hive, they all rush protectively to this side and scratch threateningly on the glass.

In fact, they were SO hyper-active that I closed the hive, let the protective activity die down, and then replenished the sugar water bottle. They were so frenetic that I thought some would try to get out the hole when I took out and inserted the bottle. I have a piece of cotton I immediately put in the hole after I remove the bottle. Then I re-insert the bottle as I remove the cotton. No time elapses between removals and insertions!

The dwindling continues and the dead bees pile up.

On the right side, bottom: the dead are one and a quarter inch high and three and a half inches long.

On the left side, bottom: the dead are one inch high and three inches long.

In the middle, bottom: the dead are one quarter to one half inch high.

Because there are so many dead bees, and they appear to be WET, dead bees, a smell of death wafts out occasionally through the vents. My 8 year old Jesse and Benjamin S. said yesterday, "Yuk! What's that terrible smell?"

I don't think it's terrible, but it is bad, sporadic and faint. That smell must be painful to my bees with their acute sense of smell, with so much of their existence based on pheromones.

"Pheromone" comes from the Greek and means "carrier of excitement." In regard to the honeybee, a pheromone is a chemical substance that is released by the bee as a liquid and is smelled by the other bees as a vapor (or a liquid if it is released near a bee). When a honeybee releases a pheromone, the other bees in the hive or around her outside the hive respond.

For instance, when she stings me, her barbed stinger comes off in me. But the muscles in the stinger continue to contract rhythmically and to pump venom into me. She, mangled, begins to die. Her stinger releases an alarm pheromone that causes other bees within a certain distance (don't know what that is) to come to her aid. They will attempt to sting me because I have her smell on me.

One of the most interesting pheromones to me is what is called the "brood recognition pheromone." These are smells that are given off by the DEVELOPING larvae and pupae in a hive. These pheromones prohibit other female worker bees in the hive from having their own "laying instinct" activated. But when there is no queen or brood in a hive and thus no pheromone to prohibit the female workers from laying eggs, they will begin to lay eggs. Now these eggs are not fertilized eggs and thus produce only drones. And we know how much the drones work in a hive! Thus the hive dies out and is called a "drone-laying hive."

There's many different pheromones in honeybees: smells when she walks over something; smells when she forages a flower; smells the queen lays down when she lays an egg and HER SMELL IN GENERAL affects the whole hive in regard to swarming, laying of eggs, social behavior, etc. She's the QUEEN, don't forget!

Let's list some of the smells (pheromones):

1. They SMELL their queen. They know where she is by her scent. They can tell a pretender because she does not smell like their queen.
2. They smell out the nectar.
3. They smell rain coming, sun coming through changes in the odor of the air. (I've observed their reaction to these.)

4. They KNOW their beekeeper by his/her smell. Maybe they know me and that is why they haven't stung me the times I was near their bee gate.
5. They smell out the pollen.
6. They smell the honey---its viscosity, its readiness for use.
7. They smell each other and know whether a bee is from their hive or not.
8. They smell the wax and the propolis.
9. They smell the presence of others---man, animals, other insects.
10. They smell any source of sugar, glucose, lactose.
11. When one of their sisters stings someone, they can smell that fact by the release of her odor when her stinger and part of her abdomen come off. They will come to her aid and will sting the person or animal or other insect.

Their sense of smell is, perhaps, paramount in regard to their faculties. This is not to demote the sense of sight or feel, but so much of their existence revolves around the sweetness of smells---the flowers they crave, the Lady they serve, the nectar they need, the wax they create. It's quite a charming way to live.

It's 40 degrees. 4:40 P.M. It was really gorgeous today---bright, high sun, no snow on the ground except eccentric old patches. It feels more like the Ides of March. When I write that, I realize how ominous is that date, the IDES OF MARCH. That was, of course, the day in 44 B.C. when Julius Caesar was assassinated. "Ides" then just meant "15th." But I'm sure the Ides of March was just another day to Julius. That morning he had sacrificed and the omens were not propitious so he had decided not to do business in the Senate that day. Plutarch who was born c. 45-50 A.D. and died in 127 A.D. gives a very detailed account of the assassination in his LIVES. (I always believe a source that is close to the time of the event rather than a 21st century "scholar" who 2,000 years later creates out of whole cloth some new-fangled position or "fact" about the ancients and their lives.) Plutarch says the assassins were men in the Roman Senate who were republicans and didn't like the fact that Caesar had declared himself Emperor. They had determined to kill him at the meeting of the Senate on the Ides of March. They all hid daggers under their robes. When Caesar got out of his litter, many rushed forward to petition him. He brushed them away. Finally, he took his seat on the Senate floor. All the conspirators huddled around him pretending to petition him. This irritated Caesar.

"...when (Caeser) saw they would not desist, he rose up violently. Tillius with both hands caught hold of his robe and pulled it off his shoulders, and Casca that stood behind him, drawing his dagger, gave him the first, but a slight wound, about the shoulder. Caesar snatching hold of the handle of the dagger and crying out aloud in Latin, 'Villain Casca, what do you do?' he then called out in Greek to his brother and bade him come and help. And by this time, finding himself struck by a great many hands, and looking around about him to see if he could force his way out, when he saw Brutus with his dagger drawn against him, he let go of Cascas' hand...and gave his body to their blows."

William Shakespeare took this account in Plutarch's Lives and made it into the play Julius Caesar. It was Shakespeare who made Caesar's last words be, "Et tu, Brutus" ("Even you, Brutus"). It is true that when Caesar saw that his good friend Brutus was one of the conspirators, he gave in to the blows. The assassination was such a haphazard one. Caesar received 23 stab

A Cosmos in my Kitchen

wounds. In the confusing melee the conspirators themselves received cuts and gashes all over their bodies. It was Shakespeare, too, who coined the phrase, "Beware the Ides of March." The Ides of March is synonymous with foreboding and death. Shows how the death of one person over 2,000 years ago can forever affect history.

From the horologist in town I got an eyepiece I attach to my glasses. I took Mother's watch to him for repair. He had this eye thing attached to his glasses and flicked it down to examine her watch. My bees! I wanted it immediately. I offered him whatever he paid for it. He sold it to me for $15.00 on the spot.

It has two small eyepieces. I attach it to my glasses. One lens is a 4 times power. With the second lens over it, it becomes a 7 power. Not only can I see the bees more minutely, I see everything more clearly. I examine virtually EVERY THING around me with these new beauties. This paper, for instance, is REALLY not at all smooth. This pen is REALLY a battering ram. This pot of parsley on the kitchen windowsill is REALLY a magnificent forest. Everything changes under magnification!

When the advance for my cookbook comes in, I'm going to buy a microscope that goes from 10 to 20 power!! Then I'll REALLY SEE!! Newman cried, "More light!" as he was dying. I'll probably be crying, "More SEEING!"

Especially with the 7th power, I can see it is still incredibly moist in the hive. Their abdomens glisten and are darker than they should be. The hairs on their heads and abdomens are sticking together. Their legs are wet. Everything is wet. Water all over the place. I don't know what to do.

"Water, water everywhere
And all the boards did shrink.
Water, water everywhere
And nor a drop to drink."   By that brilliant, opium-eating drug addict Coleridge.

There does NOT seem to be a particular place on the honey hexagon that the bees first open. I've observed opening or tapping of honey on all sides of the cell, except at the tippy top.

February 7, 1983

Last night and today we've had the biggest snowfall of the winter. It's only about 8"-9", but it's enough to close all the schools. Blake (17) and Jesse (8) are home and happy. I love it when school is closed and the children are home! I got Jesse a sled and he and Blake went out and made sled runs on Parley Lane. I took pictures of my two wonderful boys.

Then we got cozy upstairs and looked through and read parts of Samuel Taylor Coleridge's "Rime Of The Ancient Mariner." I have a Gustave Dore-illustrated book of that strong Christian poem about sin and salvation. (All three of the children and I love Gustave Dore's illustrations. I have the Coleridge book plus Dore's illustrations to Dante's <u>Divine Comedy</u>.) There in front of us is the Ancient Mariner, bedraggled and bewildered, with the huge albatross around his neck. He had killed the bird of good omen. The crew turned against him. They hung the albatross, a tangible sign of his sins, around his neck. One by one the crew on this ill-fated, damned ship die off. In one of Dore's illustrations, the Ancient Mariner wearing the incongruous bird falls back in horror as his dead and dying mates reach out to him. He symbolizes to me the fate of all of us who carry a conscious knowledge of our sin and impending doom unless we are "saved" from that fate. John Bunyan in <u>A Pilgrim's Progess</u>, another allegory like "The Rime," has Christian flee from the City Of Destruction. Bunyan, like Coleridge, has his character carry a symbolic burden on his back. I'm sure Coleridge knew <u>Pilgrim's Progress</u> down cold. That allegory of allegories was written over 100 years before Samuel T. Coleridge was born. Coleridge gave his Mariner a symbol for sin ("an albatross around my neck") that has endured as strongly as Bunyan's "burden."

"Ah! Well a-day! What evil looks
Had I from old and young!
Instead of the cross, the albatross
About my neck has hung."

When I opened the hive today, there were NO bees on this side. Very shortly, there were ten. I hope the "dwindling" hasn't gone too far!!

No sign of laying on this side. I hope she is okay and has laid the spring brood. I hope there's enough workers to midwive their births.

I'm a little worried. In the last three days the pile of dead bees on the left hand side of the bottom of the hive has developed a white substance all around them---encasing them. The dead ones in the middle and on the right don't have that---yet.

The substance is milky white and appears to be relatively solid so the dead bees are "locked" in.

It couldn't be wax. I doubt seriously if they would be in process of waxing them in? But I've read instances where they've waxed in a dead mouse to avoid putrefying.

The dead are really smelling. Could they be waxing them over? What a monstrous job it must be to encase a mouse who wandered into a wild hive, was stung to death by hundreds of bees and then had to be waxed over by thousands of bees!

The white could, also, be the by-product of decaying.

There are now 27 bees on this side. Think I'll leave the hive open and give them a little kitchen warmth. It's 30 degrees outside; 1:00 afternoon.

February 10, 1983

18 degrees F. It is a depressing time for my hive. I open it to devastation, death, dwindling. It is an over-humid environment with hundreds of bees dead and rotting on the bottom of the hive. A post-holocaust atmosphere.

Note inserted later: Have been reading about the Black Death, the Bubonic Plague that periodically ravaged Europe and the rest of the world starting in the 1340's. One out of two people in Europe died. They estimate that one out of three died worldwide. Yes, my bees are just insects and their death IS different from human death. But the atmosphere in my hive is analogous to the atmosphere that a survivor of the Plague, Agnolo di Turo, describes in his gripping short account of the ravages of the Plague in Siena, Italy:

"The mortality which was a thing horrible and cruel, began in Siena in the month of May (1348). I do not know from where came this cruelty....There are not words to describe how horrible these events have been....The infected die almost immediately. They swell beneath the armpits and in the groin, and fall over while talking. Fathers abandon their sons, wives their husbands and one brother the other...it appears that this plague can be communicated through bad breath and even by just seeing one of the infected....Those who get infected in their own house, they remove them the best way they can....No one controls anything and they do not even ring the church bells anymore. Throughout Siena, giant pits are being excavated for the multitudes of the dead and the hundreds that die every night. The bodies are thrown into these mass graves....When those ditches are full, new ditches are dug. So many have died that new pits have to be made every day. And I, Agnolo di Tura, called the Fat, have buried five of my sons with my own hands....there is no one who weeps for any of the dead, for instead everyone awaits their own impending death. Medicine and other cures do not work....I have thought so much about these events that I cannot tell the stories any longer. This is how the people lived until September (1348)....One would find that in this period of time (4 months) more people died than in twenty years or more. In Siena alone, 36,000 people have died. If you count the elderly and others, the number could be 52,000 in total. In all the boroughs, the number could be as high as 30,000 more. So it can be seen that in total the death toll may be as high as 80,000 (in 4 months). There are only about 10,000 people left in the city and those that live on are hopeless and in utter despair. They leave their homes....Gold, silver and copper lay scattered about. In the countryside, even more died....I cannot write about the cruelties that existed in the countryside: that wolves and other wild beasts eat the improperly buried....The city of Siena appeared uninhabited....everyone who survived celebrated his or her fate....Now, no one knows how to put their life back in order."

The Plague spread to Ireland in August 1348. Friar John Clyn survived and was an eyewitness:

"That disease entirely stripped vills, cities, castles and towns of inhabitants of men….(it) was so contagious that those touching the dead or even the sick were immediately infected and died…confessing and confessor were together led to the grave….(in) Dublin alone from the beginning of August right up to Christmas, fourteen thousand men (people) died."

I LOVED Ingmar Bergman's films years ago. My favorite was the Dark Ages film <u>The Seventh Seal.</u> The Dance of Death (La Danse Macabre inspired by the Great Plague) haunts me still. Dancing Death leads pope and child and king and commoner in a crooked single-file jig silhouetted against the cold Norse sky. The Black Death leveled all classes. Death doesn't play favorites.

In some way in this Fallen World there is a raison d'etre for every thing and everything. I have no trouble bowing mentally and spiritually, mote of dust that I am, to Him Who Created all, Sustains all and will in Time Redeem All. I KNOW THAT NOTHING IS LOST HERE AND IN THE AFTERLIFE IT WILL ALL BE UNDERSTOOD AND RECTIFIED. I am dumbfounded by the infinity of Luke 12:6,7 let alone the infinity of it ALL: "Are not five sparrows sold for two pennies? Yet not one of them is forgotten by God. Indeed the very hairs on your head are numbered. Don't be afraid; you are worth more than many sparrows." He knows every teeny follicle of fur on each of my bees! Evil, sin and death are all INTEGRAL parts of this Fallen World. Man's Choice caused the Fall and men's choices cause him/her to err. The Great Dance and Drama of earthly life is too great for me to fathom. (God knows, I've tried!) I'm finite, limited. He is not. I'll leave the macrocosm to Him and concentrate on the small "m."

The dark mahogany (small "m") cells. The glass with long brown rivers of resin is pocked with rocks of white

A Cosmos in my Kitchen

wax. The wood looks old, used and stained with the generations of getting and begetting. Solitary bees slowly dying, trying to rise only to fall again.

My hive has been through many lifetimes, yet it is less than one year old. When people ask me about the life expectancy of bees and I tell them it is about 6 weeks, they are shocked at the brevity. Yet I see little difference between their life and mine: conceived, carried, birthed, coddled, launched, occupied, dead. They KNOW in their genes that all is for the perpetuation of the hive. Most of us don't KNOW this. But our physical life moves inexorably and certainly toward the same ends---perpetuation and protection of the species.

The only difference between them and me is that I am WATCHING them and RECORDING how they live and die. Man's reflective, contemplative nature separates him from them. God breathed His Essence into us and we became living SOULS.

My life hurries to its end as speedily and inexorably as does hers. There are even insects whose entire lifetime consists of 24 hours. I think the fruit fly is one. That is why they took them aboard space ships.

If one day of my bee's life would equal one year of my life, when she was 6 weeks old and ready to die, I would be 42 years old. Many die at 42 or younger. This primitive ratio amply illustrates time's relativity. The Book says, "One day is with the Lord as a thousand years and a thousand years as one day." So to the Lord, my 42 years is 15,330 years. We're into a place where Time is so relevant, it is not relevant. Truth with a capital "T" is paradox again as I figured out in college.

The white stuff observed three days ago has increased in mass and visibility. Still confined to the lower left bottom and almost thoroughly encasing the dead, wet, rotting bees. Snow white extends from a sixteenth of an inch crust down through the layers of dead bees. What is it?

February 12, 1983

Yesterday and last night we had a terrible blizzard. The radio and TV have already dubbed it The Blizzard of '83. We got 18-20" of snow. There are 5 ft. drifts and I sink up to my knees when I walk from the main house to the carriage house. It is 18 degrees and during the blizzard, the winds were 30-50 mph. We were inside---cozy, warm, fed. How blessed! Bless and care for those who are homeless in this, Lord!

This morning when I opened the hive, there were 3-4 bees so cold they were not able to move. I turned on the overhead light and put the heater near the hive. Gradually they have thawed out. Now there are 26 bees on the honeycomb and three times that number on the brood comb.

They have dwindled so drastically in the last month that I sometimes think there are only 10-12 bees left. So it's good to see so many more than that number.

The snow has totally blocked the bee gate, but today there was a strong sun and I'm sure the next days will melt that blockage.

Each day more are dead in the pile at the bottom of the hive. It saddens me and I worry that they will not be able to carry out all the dead. It will be a Herculean task. We'll see.

The milk-white substance encasing the bees in the lower left is now beginning to be streaked in bright orange. I do have a theory re: the orange color---perhaps it comes from the color on their abdomens because the bright orange is located where the orange on their abdomens is located. It's interesting really how the bees, their hives, their products and by-products are only several colors.

WHITE:

1. The wax.
2. The honeycomb.
3. The brood comb before use.
4. The milky substance I've observed.
5. Light, sun.
6. The 1,000's of tiny, white crystals that they store and stash in the area between the honey and brood combs. I don't think they are bits of wax.
7. The white the honey cell turns before being tapped. As has been observed, chronicled and verified in these notes, cells ready to be tapped by the bees in my hive take on a pure white color.
8. The white glistening larvae.

BROWN (and shades: tan, reddish brown)

1. Parts of their anatomy.
2. The brood comb after use.
3. Propolis.
4. The color the wood in the hive turns after they have inhabited the hive.

GOLDEN (and shades: clear, yellow):

1. The nectar.

2. The ripe honey.
3. The pollen.
4. Parts of their anatomy.
5. The eyes of the drone.
6. The sun.

NOTE: Of course, there are flowers. And, I must say, MANY of their favorite foraging flowers are White and Yellow.

BLACK:

1. Parts of their anatomy.
2. Night, darkness.

As can be seen, the bee's life is white, brown, black and golden. (Of course, there's the striking red color of her proboscis!) She who is associated intimately with and genetically tied to all the flower colors of the rainbow is in and of herself rather monochromatic. The only primary color associated with her life is yellow---the color of honey, pollen and light. But when I look into her world, into the hive, the color is golden brown, not bright yellow.

The true primary colors are red, yellow and blue. Just think. From the mixture of those three primary colors come all other colors. Yellow mixed with blue becomes green, etc.

I've read that honeybees can't see red. How do they know? I know it has something to do with ultraviolet rays. But you couldn't know unless you were a honeybee. And I'm sure they don't have words for the colors. They probably have instinct for the colors. Or maybe since they are so pheromone-oriented, they identify the flowers through smell. The most amazing part of the bee to me is the eye. I should say eyes because they have three small eyes on top of their head and then two huge eyes---one on each side of the head. These huge eyes are EACH composed of thousands of lenses (some say up to 8,000)! That would be like me having two eyes EACH of which has thousands of lenses in them! I would see a constant mosaic. Thousands of human eyes in ONE of their eyes. Of course, we'll never be able to understand what it is like to see the world through the eyes of a honeybee. How could we? If I could for 4 seconds see the world through the eyes of a honeybee, it would probably leave me crazy or on my knees!

Colors. Some honeys are brown. Raw honey is the best because they haven't heated and centrifuged all the wax and propolis and bee bits away. Raw honey is definitely brownish. And it's the best for your health.

I read there was a blue honey in North Carolina, but they couldn't determine the nectar source. And, of course, the fabled white honey made 2,000 years ago from the wild thyme that grew on Mt. Hymettus in Greece. I'd love to see, let alone taste, white honey!

Most of the honeys we get in the supermarkets are a mixture of various nectars. The clovers are definitely the bees' primary nectar source. Clover honey looks beautiful on the shelves. It's the color of what we perceive as "honey." Clear, golden. Thick and delicious on hot biscuits with butter.

Nowadays, people don't give honey to their little babies. But I did. I gave all three of them honey and water for about 6 weeks when they were newborns. Even before I got into bees, I felt that honey was a good thing. Plus, honey is a soporific, a natural, non-addictive (!) aid to sleep.

There are, however, some honeys that are poisonous (which is the reason no one gives their babies honey anymore---it takes one case only and the whole world swears off something!). The honeybee normally collects from many, many sources. If, however, she only has ONE source and forages JUST THAT PLANT, some honeys can be poisonous. E.g---black nightshade, locoweed, mountain laurel, California buckeye, azalea, rhododendron.

When I took Greek in college, one of the books we read was Xenophon's <u>Anabasis</u> which means in Greek "the march up." Xenophon lived c. 400 to middle 300's B.C. The <u>Anabasis</u> tells his story of how he and 10,000 Greek mercenaries marched to Babylon to aid Cyrus in overthrowing his brother. It didn't happen and he and the others, exhausted, had to "march back" (called the "katabasis") to the Black Sea. Well, the famous mass honey-poisoning story is in that ancient book. The Greeks were camped in a woods near the sea. The woods were covered with wild rhododendrons. There were many beehives in the trees. The soldiers were ecstatic because honey was man's ONLY SWEET until the sugar cane was introduced into Europe in the 1200's A.D. Here's how Xenophon describes the poisoning:

"All the soldiers who ate of the honeycombs lost their senses and were seized with vomiting and purging, none of them being able to stand on their legs. Those who ate but a little were like men very drunk, and those who ate much, like madmen and some like dying persons. In this condition great numbers lay on the ground, as if there had been a defeat, and the sorrow was general. The next day none of them died, but recovered their senses about the same hour they were seized. And the third day they got up as if they had taken a strong potion."

A Cosmos in my Kitchen

SO---over 2,000 years ago a Greek army DID get very sick from eating rhododendron-nectar-laced honey. That bush was the sole source of the nectar. Very, very rare occurrence.

February 13, 1983

2:10 P.M. 36 degrees and sunny. Almost every time I open the hive there are new things. I can't believe how variegated these bees' lives are.

Today on the inside bottom of six empty honeycomb cells there are deposits of brown "goop." As I look closer, I see it is, also, inside some of the empty brood comb cells. I can think of two possibilities: propolis or poop. It would make sense to put either one in the empty cells until it could be redistributed or taken outside. They are stashing whatever it is: way in the back of the cell; in the front and interior of the front; on the outside rims of the cells, like surface storage vs. interior storage.

Nine out of ten of the bees still display the effects of the humidity: orangish/black abdomens vs. honey/brownish PLUS the hairs on the thorax are matted together vs. light, erect and fluffy. I can see all this very well with my magnification. Makes me wonder where the occasional "normal-looking" bee has been that she doesn't show these effects.

The strong sun has melted the snow at the entrance to the bee gate, so it is clear.

I love the way they preen, clean and hone their proboscises (their mouths). When I was young, I thought a "proboscis" was a "rear end." My mother used to call our rear ends our "ssa, ssa." It wasn't until I was much older that I realized "ssa, ssa" was just "ass, ass" backward. Mother who is very proper didn't know that, I'm sure.

Anyway a proboscis is a mouth and the honeybee's mouth is very complicated. It's a long, tubular creation that extends out from the bottom of the head when needed to ingest nectar, propolis, etc. Also, it has a further feathery apparatus like a tongue that extends from the tube. I've observed that this acts like a vacuum cleaner. The hairy extension will swing back and forth across the honey in a circular or a lateral motion lapping up everything in its path. Very feathery and efficient. When not in use, the hairy tongue as well as the proboscis fold down and tuck under the bee's "chin" (for want of a better word).

When they preen, clean and hone their proboscises, they brace themselves against the hive using their 4 anterior legs. (Bees are insects and have 3 pairs of legs---6 total.) They shoot out their red proboscis. Then they take the 2 front legs and groom the proboscis starting from the top near the mouth. Clasping the proboscis with one leg on each side of it, they slide the two legs all the way down to the tip of the proboscis and then repeat this grooming over and over. The usual amount of times is 8. They seem to have to keep their "mouths" clean and preened. Like brushing your teeth. That's not a perfect analogy because you can still eat if you never brush your teeth. The honeybee's proboscis has to be well-groomed as it is vital not only for her life, but for the life of the whole hive and developing brood.

Ever since I've had the hive (over 10 months now), I've had a problem with the bottle feeder. You're supposed to hook the feeder on with a spring, but I found that emptied the sugar water into the bottom of the hive because the angle wasn't right. So I abandoned the spring and let the bottle lay over against the window.

I found that angle worked, but now I hold the bottle in place at that angle with common clay. It's perfect.

I came to that solution for the bottle feeder as a result of another problem. There is a plastic casing at the bottom of the bottle around the metal tube feeder which ideally is supposed to close the hole where the feeder goes in so that the bees can't get out into my kitchen. Well, the plastic cap got stuck INSIDE the hive (rather than half in and half out). Don't ask me how it got inside the hive. It did. So there was the hole and the little bees could squeeze out and get into my kitchen. So I ran for Jesse's clay and found to my delight that the clay that plugged up the space on either side of the feeder also kept the feeder in place at the right angle for bee consumption. It's not infrequent that one solution solves two problems.

I can't believe how fearful and upset I get when the sugar water at the manual's angle starts emptying into the hive. Shows how invested I am! Of course, my bees just shoot out their nectar tongues and within an hour the floor is clean.

I just put a full jar of sugar water to the hive. We'll see how long it takes them to empty it.

I have been leaving the hive open for 4-5 hrs. each day recently. This side is always empty when I open it up to observe them. Almost immediately, the bees come. I think they like the added light and warmth of the kitchen. I hope I'm doing them a favor.

When I smell inside the air vents on the lower brood comb, the smell is death and decay. But the smell from the vents up on the honeycomb is still cinnamony, clean and exotic.

I hate to mention the pile of dead and dying bees on the bottom of the hive. They're all drenched and discolored with moisture. The dead bees outside the hive between the screen and storm window on the sill look like normal dead bees. In times past I have seen the bees go out there to die. Like elephants and sharks, they go to a preordained graveyard. They are fluffy and honey-colored. They are just dying. That's all. But the dead hive bees are damp, drenched and soggy.

The ones who stay in to die wade haltingly over the dead. They try vainly to prolong their activity. They fall, roll around on their backs, become entangled in the feelers and feet of the dead. There are 5 layers of dead bees in the bottom of the hive. In the lower right hand corner there are 9 layers of dead bees.

Years ago I went to Myrtle Beach, South Carolina with Jack and Keith and Carole. The big tourist thing was to gather petrified sharks' teeth on the beach and have the teeth made into bracelets. The waves whoosh in and the moon drags them out again. I would quickly look to see if it had washed up any petrified sharks' teeth. Oh, there's one! There's another! And I would run through the shallow water and capture it in a clump of wet sand before the next wave came in and washed it out again. The locals say that many miles offshore there is a prehistoric graveyard of sharks. That sharks knew eons ago that this was the spot of ocean where they had to go to die. When it came their time to die, they swam from India or Maine or wherever and died there. Ah, here's where I'm going to die. And circling, circling, eyeing the dead below. Knowing, knowing that I will be next. All that is left now after eons is their teeth. And they wash ashore at Myrtle Beach. And the-eons-later-people (us) gather them, make jewelry out of them and tell people: "Oh, these are petrified shark's teeth. I gathered them at

Myrtle Beach." Interesting that all's that's left of most people/animals after a long time is their teeth. Think of how many missing people have been identified by a tooth.

The queen!! I just saw her for the first time in 6 months! She's on the brood comb with two attendants. She's inspecting the cells.

She's not at ALL wet or damp. Her thorax is fuzzy and tan. Her abdomen is over a half inch long and honey-colored all over. She's beautiful. It's been so long since I've seen her, I almost didn't recognize her.

She's had her head in one cell now for 2 minutes with one attendant on her right side. On the entire brood comb there are only 6 live bees scattered about. There are 10-12 dying in the pile at the bottom. Three bees are near her. The rest are scattered. She's picked the ONE place on the brood comb (top center) that has a 2" smear of propolis and wax on the glass. Thus she is partially hidden from me now. It's almost as if she picked that one spot where she would be most obscured. Darn!

She is now communicating via feelers most intensely with an attendant. There are now 3 in a semi-circle around her. One at her abdomen touching her with her feelers. Now here comes another and another---3. This is an information gathering and check of the territory tour.

I guess when the dwindling, dying gets this drastic, her entourage drops from 12-15 to 3-5.

One attendant has been grooming her abdomen with her proboscis for over 5 minutes. She remains immobile, allowing this. Maybe the fact I've had the hive open for 6 hrs. today has lured her into the warmth?

Thought: Always the stroking of the queen with their feelers and proboscises. Is it possible that the attendants help in the egg laying? I mean, they KNOW via their feelers the state of readiness of an egg to be laid PLUS the stroking helps to formulate, regulate, facilitate the egg laying. What I'm saying is---she is not autonomous. EGG-LAYING IS A COLLABORATIVE EFFORT. Maybe she could not lay consistently without her retinue. Are there observations on this?

Everything else about the honeybee's life is interdependent. Why not this most vital, sine qua non aspect? A more scientific observer than I would have to verify this, but it sure should be examined.

Theory: QUEEN'S ATTENDANTS AS CO-LABORERS. The attendants are not just there to guard, protect, feed and cater to her. They are co-laborers WITH her in the reproductive process of the hive! I feel in my bones this is the truth. I have always sensed their ACTIVE participation as I watched her lay.

February 15, 1983

Well, I got 4 small packs of silca gel. The man in the camera store was nice enough to give me the 4 packs out of camera packages. I've got each one tacked in front of each vent to try to absorb some of the moisture in the hive.

I waded through 4 ft. drifts and climbed up to the bee gate outside. I scooped out as many dead, soaked bees as I could. Again they had filled up and stopped up the bee gate. I came back in and got an iced tea spoon, went out and scooped as far up inside as I could. Then scooped as many as I could from the bottom of the windowsill outside the bee gate. Thousands of little ones peppered the salted snow.

The Lord knows why I'm trying. I may be doing the wrong things to help them, but I'm well-intentioned and, yes, I know the road to hell is paved with good intentions.

They do respond to the hive being open and to the light. They are lethargic, dispirited when I open the hive, but they become more animated with the light and warmth. I am the same way.

Even her Majesty comes out now each day when the hive is open. She's alone, enters from the upper brood comb right or left and slowly, as is her wont, makes her way over the empty, brown cells to the center of the hive. Then she goes to the very top of the brood comb and stays there---away from the dead and dying on the bottom. But she is today and was yesterday totally alone without even one attendant. She is dry, unlike the others. She does get wet on her back when she presses between the moist glass and the wood. She rests at the top of the brood comb.

Oh, what will be the resolution of all this death, decay and dampness?!

February 16, 1983

MY MOTHER'S BIRTHDAY. SHE IS AND HAS BEEN MY QUEEN! I thank God He gave me to her and to my Dad.

Today the queen was down on the bottom of the hive in the pile with the dead. She did manage to climb out.

I can't write. I'm heartsick.

Took silica-gel away and will let nature take its course. Will leave hive closed, light off and PRAY.

February 17, 1983

I think all is lost.

One or two are half-alive at the bottom of the hive. Tumbleweed everywhere. Can't write anymore. Too sick.

March 2, 1983

Each day I have opened the hive. NO life. Death and decay everywhere.

March 7, 1983

Still no sign of life except: I've marked the sugar water bottle and it has gone down over 1" in the last six days. Either it is dripping or someone or ones are taking some. I pray the latter, but when I see the dead, molding, rotting bees and I see the condensation on the glass all turned to a white film and I see white mold beginning to creep into the brood cells---it is hard to hope.

I sit for 10-15 minutes with my ear to the air vent. I imagine I hear buzzing, but I think it is only the echo of passing cars from outside coming through the bee gate. I hold my nose against the stench of the rotting bodies.

March 15, 1983

The Ides of March---Beware.

I was in the yard clearing away winter debris from a garden. It was 53 degrees, sunny. A honeybee came down and buzzed me. I told her to let me see her. She lit by my hand and we touched each other!

I ran to my hive, but no sign of life. Maybe she was a descendant of one of my swarms and knew my scent.

March 16, 1983

52 degrees. Sunny.

Honeybees are about in the gardens. I ran to the hive. No sign of life.

April 5, 1983

They have all died. The hive is filled with green and white mold.

Lena and I broke the glass today and went in to clean it out. The feeling was of an insect Dachau---dead, emaciated, decomposing bodies. Filth everywhere. We cleaned them all out.

God bless Steve. On Saturday, April 2, he took the hive out of the window, put it in the garage and with much wrenching took the top of the hive off and extracted the honeycombs. They had propolised the top shut and it was a horrendous job getting it unstuck. Steve's hands were cut and bloody from the job. Now the hive has human blood all over it.

I looked for the queen among the dead but couldn't find her.

No more talk of the dead. Now I want happy things. And there are some. I took one of the honey-laden combs, cleaned off some of the mold from the wax and tasted the very gold honey. It tasted light with a small hint of fermentation. I took the comb outside and put it in the Heath, Heather and Thyme Garden for the bees.

I've always been fascinated that the word "honeymoon" seems to refer to the old Viking practice of the couple drinking metheglin (honey wine) every day for the first month of their marriage. I get the "honey" part. Does the "moon" part mean the 30 or so days of the month? Probably not, because that's a solar month. It probably means they drank for a lunar month of 28 days. And "luna" is the Latin word for "moon." "Luna" is, also, hidden in the word "lunatic."

If you wanted to, you could say that the combination of mead drinking for days on end and the forced togetherness and the fact that many of those women were abducted and not willing brides produces--- "lunacy." As any emergency room worker knows, there's always more ambulances during a full moon than at any other time of the month. So the "moon" part of "honeymoon" maybe is the lunacy part? Maybe I'm "loony" (cute form of "luna") even trying to figure this all out? Or maybe "honeymoon" just means that the couple had one "sweet time."

I'm writing beside the honeycomb sitting on thyme (I like the idea of sitting and walking all over time), lemon thyme with the carraway thyme trailing down the stones under my pant leg. Ten bees have found this comb, have forsaken the nectar of the tiny spring white flowers of the erica nearby and are eating the matured product. I thought they would eschew the part of the comb with the mold on it and go first to the part I had cleaned off. But they're all over: six on the clean part and four on the moldy part. They are sipping away to their hearts' content.

It is fitting that all the work from my bees should go to other bees. I'm going to keep some for us, but leave a tithe, a gift, an offering of it to them.

10 minutes later. Sixteen bees and one fly are on the honeycomb. The fly walks slowly about with his proboscis extended. As far as I can detect, he isn't feeding. He's just amazed, stunned at this new source.

1 minute later. Twenty-one honeybees. I am 3" away from them. They buzz and fly and land on me as if I were part of the comb. They know who I am.

I do see now that there is a preference for the top part of the comb that I cleared off. They are emptying the tapped cells first. They leave the unbroken ones alone. More come every minute. They leave to take home the honey (?) and then hurry back.

A little black ant has entered the picture---teeny and bewildered by all the good food.

April 19, 1983

Big Bee Day. Hundreds of honeybees picked the three brood combs (six if you count both sides) clean of every smidgen of honey in just 7 hours! It was 62 degrees, sunny and they came in force. I sat among them and they were friendly. They would become laden, drunk, sticky and would have to clean, unsticky (honey is sticky) themselves in several ways:

1. They used a pile of dry, dead leaves. They would rub against them until they were clean enough to fly.
2. They used the many, tiny dry stones of the driveway. They would agitate in them until enough of the honey came off of their wings to permit flight.
3. They would groom one another until they were able to fly.
4. I had the combs in a large turkey pan. They would go to the rim until they had groomed themselves enough to fly. The ENTIRE RIM OF THE PAN is now coated with wax shed from their bodies.

So even though it has been a surprise feast for them, it is a sticky, waxy business, this getting of free honey from the comb. Sticky and waxy for me and for them.

On April 5, I cut off and threw away all the tops of the honeycombs and harvested some of the honey. The tops were coated with green and white mold. I scraped out the wax and honey into a large kettle. I melted it all on low heat until the honey and wax melanged. Then I let it cool. The wax rose to the top and hardened. I lifted off the wax. I strained the honey through a gauze-like cloth into jars. It was very time-consuming. I got two quarts from that tiny hive and that's after they had lived off the stores all winter!

State of upper honeycomb stores when I harvested honey from hive:

A Cosmos in my Kitchen

The side toward me: the bottom was empty, but the top was full. On the other side, it was untapped.

The middle comb: the bottom was empty, but the top was full. On the other side, it was untapped.

The comb toward the outside: BOTH sides were full.

This throws my honey consumption calculations to the wind. They consume MUCH LESS per day. The economy of consumption is phenomenal!

After I strained the honey into a quart jar and 2 pint jars, I melted the wax. It was filled with "things." I skimmed off the clear wax and capped the jars of honey with it---nature's paraffin. The color of the wax is clear white and it is soft as a baby's bottom.

I put two cinnamon sticks and two slices of lemon in one of the pint jars. Capped, it looks lovely.

"If you a cook of note would be,
Use honey in your recipe." Harriett M. Grace (1947)

I probably will NOT use this honey instead of sugar in my recipes, but I will eat it raw and will drizzle it over hot buttered muffins.

As I was processing the honey, I kept tasting it. It tasted slightly fermented. But now that it has set for several days, it tastes like good honey. The color is brownish gold, not golden. It was an experience to harvest and to put it up. I wept for their work and their deaths. I never intended them to work to give me anything.

I just read the previous entry. "To put up" means, of course, "to can" something. Which means to prepare fruit, vegetables, relishes, meats and put them in jars for preservation and future delight. I've looked all over and can't find where I read about this woman in Pompeii in 79 A.D. who preserved figs in clay jars. Of course, on August 24, 79 A.D. Mt. Vesuvius erupted and she and her preserved figs along with the whole town and its inhabitants were totally destroyed and "preserved" in tact under more than 60 feet of volcanic ash. In the late 19th century archaeologists discovered her kitchen and there were the jars. They broke one open. The figs still looked good. "I'll taste one," said a bold scientist. He did. They were still fresh!! I often think of that woman. If she had lived a million years, she would never have been able to imagine that those figs she put up so casually and lovingly on say, an Ides of "July" (named for Julius Caesar), would still be good and would be tasted by a person who lived 1,800 years later!! I love those historic ironies! And "they" say we've only been "canning" since the early 1800's A.D.!

I looked it up. It IS true that we've only be putting things in glass jars, in tins and in wrought-iron canisters ("can" is short for "canister") since the early 1800's. Seems the impetus for canning was WAR. Lots of wars and rumors of wars mean lots of soldiers who need to be fed. How to get fresh, not spoiled, food to all these fighters? Put the ravioli, the corned beef, the pork and beans in tins and ship them to wherever. The French pioneered canning in glass jars (breakable) in the early 1800's. Then the British improved on the system by putting food in wrought-iron cans (not breakable). Within a decade America was tinning food. The new canned food was considered the best thing since sliced bread (the world would have to wait another 120 years until 1928 for that innovation). Everybody loved those tinned oysters, coq au vin and pears. But the article I read emphasized that the driving force behind all canning and improvements in canning was war. Tell that to my mother who canned fruit from our fruit trees until she was blue in the face. Tell that to my grandmother who taught me how to can those wonderful peaches. Tell that to the Roman woman in Pompeii placing plump figs inside sturdy clay jars on that sunny afternoon in July of 79 A.D.

Today my bees' offspring, perhaps, harvested what was left in the combs. So their work and product was to avail.

Went to Ed and Nita's. Ed examined the brood comb and said it was all right to reuse. Nita's gardens are so beautiful! That's really how they got into beekeeping. Nita read that bees helped your garden become more beautiful by cross-pollination. At one time Ed had over 200 hives. Not all of them were on his property there in Wilton. But if you think how much honey he can get from those 200 hives: 100 lbs. of honey per hive in an average year equals 10 tons of honey a year! He doesn't, I'm sure, harvest that much. But locally he makes a little on his Wilton Gold honey (a brown honey). Ed's 200 hives contain over 10 million bees.

Note inserted later: Was just reading up on cross-pollination. It's amazing that all these words and phrases we hear and learn in school about the "birds and the bees" have very little real meaning for most of us. At least they didn't for me at that time. I'm looking at a cross diagram of a flower. The STAMEN is the MALE part on the outside of the flower that holds the pollen SEEDS. The PISTIL is the FEMALE ORGAN hidden deep in the bowels of the flower. When the honeybee visits one flower, she picks up its pollen on her furry coat. As she probes for nectar in the next flower, some of the pollen from the previous flower is sprinkled on that flower. Now---if the pollen is COMPATIBLE with that next flower, the pollen seed sends out a pollen TUBE that goes all the way down in the flower TO THE OVULE. It is FERTILIZED! If the pollen is not compatible, it sends out a pollen tube anyway. But the pollen tube doesn't go very far into the flower. This is called an "aborted pollen tube." Cross-pollination MUST occur for a flower to produce fruit. It's like human sex. Male seed. Female ovary. Tube. Conception. Fruit of conception is a baby. The phrase "Fruit of the womb" does, indeed, imply cross-fertilization (cross-pollination). I have to put this in such elementary terms because I'm not a science/math type. NOW I understand on a basic level why they need to transport all those beehives to Maine when the blueberries are in bloom or to California when the almond trees are in bloom or to Florida when the orange trees are in bloom. These crops would undoubtedly produce some berries, nuts and fruits. BUT the mass amount of imported honeybees flying constantly from flower to flower ensure that there will be an AMPLE harvest of broccoli, zucchini, tomatoes. We are literally dependent on these tiny insects for the fruits and vegetables we eat! And for the gardens we love and the flowers we admire and pick to put in pretty vases on our kitchen tables!

I always get my propolis from Ed. Some make it into tablets or gel caps, but I get it as a tincture suspended in alcohol from Ed. I use it for many things. Am convinced that propolis will one day be "discovered" by the medical profession as helpful in allaying skin diseases and stomach problems as well as other things.

Still cleaning out the hive. Right now I'm taking out the brood that died in their birth cells. Almost all of them were fully developed.

Some observations:

1. As I extracted the dead bees from their capped brood cells, I find that ALL of them are in the cell like this:

Their head is at the top of the cell, abdomen at bottom---without exception. This explains a very early observation: when they are born, they immediately head UP to the top of the brood comb. She is so positioned in the cell that she crawls out with her back toward the glass and her head pointed to the top of the comb. She's positioned toward the food.

2. When they were extracted from the cells, every one had her PROBOSCIS OUT. Now, does this happen at brood death OR are they in the cell with their proboscis extended so that at the time of birth, they will be able to eat right away OR...?

April 22, 1983

We hived a new package of bees today!

I think I over-drenched them with sugar water for two reasons:

1. I was scared and wanted them quiescent.
2. Jesse, 7, was right there. He insisted on being at the hiving. He said, "I'm not afraid of them, Mom. Please let me stay. You and I are a team!" What Mother could resist that last phrase! He put my Mom's plastic rain scarf over his face. I sprayed them all the more with sugar water because my little one was there.

This time Steve had my lime-green see-through nightgown over his head. I was the only one with a bee hat (on loan from Nita). The three of us were a sight: plastic rain scarf, nightgown and bee hat.

Watching the hiving were Lena, Maggie (her friend from Sweden), my mother and Blake. It went off without a hitch. (Nothing like the drama of the first hiving.) It was a cloudy day. I put them in the garage for the normal process of the bees going into the hive from the package.

April 28, 1983

Today my dear father would have been 75 yrs. old. He died March 12, 1979.

It was in the 80's today and sunny. On the day he was born in 1908 in Chillicothe, Ohio there was a terrible snowstorm. I'm sorry he isn't here to see my bees. He would have considered me crazy, but would have been proud of my hobby and, I think, he would have found them interesting.

If the hiving was routine, the post-hiving was not! It rained the afternoon and evening of the hiving. It rained the next day and the next. The hive sat in the garage. The bees ran in very slowly. Finally two days after the hiving, I hit the top of the package and knocked down a swarm-like cluster toward the hive hole. Some went in. Tuesday I shook as many as I could into the hive and closed it up. Steve carried it into the kitchen and installed it.

There were many dead on the floor of the hive. Lots of condensation on the glass. Oh, no. I was worried.

The next day, Wed., it was in the 70's and sunny. When I awoke, I ran to my bedroom window in anticipation of seeing all my bees outside the bee gate gamboling.

No.

7:00 A.M. and sunny. Took Jesse to school.

Observed at 8:30 and they were in a cluster from the brood comb center down to the bottom.

Bad sign. No in and out. All clustered together. Silent.

After all these hours of observation, I knew something was wrong.

9:05 A.M. Came back and no activity. I began to cry. ("Cry baby, Cry baby, Hanging on a bull's tail...") The bee gate was clogged. They couldn't get out.

Steve to the rescue. He takes the hive out of the window. He bends a coat hanger and goes into the bee gate. It can't make the turn up into the hive, but it comes out with some dead, wet bees on it.

Ominous.

He tries a plumber's snake. It won't make the turn. Tries a piece of plastic. It bends. Finally, he cuts up one of Jesse's Hot Wheels tracks---hard, flexible plastic. Forces it up into the hive. Jiggles it and pulls it out. Out come 4 live bees with it. HOORAY!

Sticks the cotton back in the hole. The bee gate's unplugged now!! He carries it back to the window, installs it, secures it, tapes the window so the bees can't get in the kitchen. Sandy climbs up the wobbly stool and pulls out the cotton. I check all day and all night and am not sure if I see more than the same 6-10 bees coming and going out of the gate. I'm still worried and praying.

I notice there is some cotton residue stuck near the bee gate. They are having trouble negotiating it. I got a ski pole. Wet a paper towel. Wrapped it around the end of the pole. Secured the towel with a rubber band. Then I went through the dining room window to the bee gate and rubbed away the cotton.

Today it is in the 80's and sunny. They are out in full force. Praise God.

Now I can get down to serious observation again.

1. There are several different kinds of bees in this package. My others were all the same: typical markings, black legs, etc. A few of these have RED legs and are light tan and honey-colored all over. I have seen three drones like this. Also, some of the workers have golden reddish legs and antennae. They all work as one. I'm watching one of the new "albino"(I so designate them because they are lighter than the normal Italian honeybee strain) feed one of the regular bees now. Also, some of the workers are very small compared to the others. More on the differences later.
2. Noticed they are using some of the brood cells to store parts of dead bees---wings, heads. Interesting.
3. The bottom of the honeycomb is spic and span. Have seen them pack two dead drones underneath the comb. One worker used the LENGTH of her body to force his body under the comb as another worker pulled the drone by his wing from the other side.

There are loads of dead bees in the bottom of the hive. They cleared 5" of dead ones away today on the right near the queen cage. Tonight the space is littered with dead bees again. Wonder why so many are dying---much more than last year. Could it be the different strains together?

The smell is lovely from the hive---that spicy propolis smell. I guess there was a lot of propolis left in the hive even though I used my hive tool to scrap off all I could.

If I leave the wood off the observation glass, the condensation goes away. If I put the wood back on, the condensation collects. Hive design problem? I'll experiment more with this and see if it could help solve the water problem.

They are going through about one and a half bottles of sugar water a day.

The workers are laying new honeycomb and I notice they are patching up some of the old brood comb. Their little bodies are sticking out of each cell. Masses of them are glued to other sections of the comb doing wax patchwork.

April 30, 1983

Lots of dead bees in the bottom of the hive---bad sign.

There are so many dead that they can't get them all out. So they have done an ingenious thing. They have cut off and pushed out through the air vents piles and piles of LEGS! Here on the sill in the kitchen and falling onto the kitchen floor are bee legs! Plus they have pushed legs and antennae up through the hole into the sugar water bottle. Bee parts float around in the syrup. Ingenious. The first batch of bees never disposed of body parts in these ways. New things to observe all the time!

I do hope these little ones will be okay with all the water and the dead?

June 6, 1983

Well, the second batch of bees went the way of all flesh. Moisture built up and up and they couldn't absorb it. Steve had the hive in and out of the window three times as we tried to solve the water problem. Last time I left them in the garage hoping that would do something. We took off the bottom of the bee gate. They hauled out the dead and I swept them away, but the living couldn't keep up with the dead and the dead won.

Ed and I have hypothesized ad infinitum re: cause of death of the two hives. He thinks it's a mystery. I KNOW IT IS THE EXCESSIVE MOISTURE THAT BUILDS UP. I'M SURE IT IS A FAILURE IN THE DESIGN OF THIS INDOOR HIVE OR THERE ARE TOO MANY BEES FOR THIS AMOUNT OF SPACE. But what CAUSES the moisture to accumulate, that's the rub.

The second doomed batch was initially put in a clean, dry hive. I installed brand new honeycomb foundation. Ed said to leave the old brood combs, so I did. Maybe that contributed to the problem?

They had used the bottom half of the old brood combs to get wax to rebuild some of the damaged brood combs and, also, to cap the developing larvae. When I went into the hive, the larvae were laid well---in a circle from top of brood comb to bottom and nicely left to right. The larvae had OLD brown wax caps. They didn't lay new wax for them. Because there were so few of the living, they used the resources at hand rather than create new ones.

Also, loads of workers were dead IN the brood cells, bottoms up. They had died working!!

A Cosmos in my Kitchen

What caused this? I've always been interested in mysteries and knotty problems. When I was in 5th grade in Alliance, Ohio, Miss Reed said if anyone wanted to be famous they should decipher the Mayan hieroglyphs. I immediately became interested in the Mayans. My dear, always encouraging Mother got me Sylvanus Griswold Morley's scholarly tome The Ancient Maya for Christmas. I've studied them ever since. Still haven't deciphered the hieroglyphs or found out why they abandoned their cities in Guatemala and Honduras and moved to the Yucatan Peninsula. But I did go to the Yucatan for a summer and studied the ruins.

In 1956 when the movie Anastasia came out, I was fascinated by that mystery. Did the Czar's daughter really survive the massacre of her Romanov family and live to tell her story? I loved Thor Heyerdahl's book Aku Aku about the stone statues on Easter Island. Who made those gigantic silent sentinels looking out over the Pacific? After I became a Christian, I researched what happened to the northern Ten Tribes of Israel after they went into the Assyrian Captivity in c. 740 B.C. Those ten tribes, unlike the 2 tribes of southern Israel, never returned. They just dropped off the face of the earth. And who built Stonehenge? Some say it was the Ten Tribes! How was the Great Pyramid at Gizeh constructed? We couldn't build it now. And how about Bigfoot? Many Mt. Everest climbers and natives have claimed to see the Yeti, as they call him. Who was Jack the Ripper? And how did the negative image get on the Shroud of Turin? What exactly is the Antikythera Mechanism? I'm scratching around trying to find some information on this ancient artifact. It was found by sponge divers in a wreck off the coast of the island of Antikythera in 1900. Archaeologists and other experts can't figure out what it is. It's a bronze contraption with many gears and inscriptions. Way too sophisticated for its date---80 B.C. Some have said it is an astrolabe for doing astronomical computations.

Others have said it is a very sophisticated clock. Either of those two hypotheses cast to the winds our thinking about the ancients. In Europe they didn't have this type of mechanism until the 1700-1800's---almost 2,000 years after someone made the Antikythera Mechanism! What caused the blast in Tunguska? All that we know is that a massive event took place in Siberia near the Tunguska River in the early morning of June 30, 1908. Some thing, stronger than a nuclear bomb, exploded in the air and felled over 800 square miles of trees. Fortunately, it was in a sparsely populated part of the globe. Strangely, no one investigated the explosion until 1921. (The Russian Revolution and World War I intervened.) There were ear-witness reports to a series of successive explosions. They interviewed a few eyewitnesses who described a glowing object hurtling to earth. All described the fact that the earth shook and they heard many, many explosive sounds. No one has found a crater. Whatever it was exploded ABOVE the earth. What happened? I'm constantly finding new things to explore. Have to explore my bed right now. It's about 3:00 in the morning.

Note inserted much later: Since I mentioned it, there has been a lot of research into the Antikythera Mechanism. Some believe it is an analog computer!! Some say it is possible that it is designed on heliocentric principles rather than geocentric. That would mean that in 80 B.C. someone KNEW the earth revolved around the sun! A reconstruction of the Mechanism is in the American Computer Museum in Bozeman, Montana.

June 11, 1983---5 days after the death of the hive.

Steve and I do the auction every year at the Yankee Peddler Fair at Jesse Lee Church. When it was over today at 4:00, Steve and I went down to Ed's and picked up a "nuc" (nucleus) hive. In that hive are:

1. A queen.
2. c. 5,000 bees (vs. 10,000).
3. Three NEW brood combs and the middle one is already filled with capped brood.
4. The same honeycombs which were new, except he gave me a middle one filled with capped honey.

So I have queen, workers, brood and honey---a clean, already working hive.

The difference between the kinds of hives I have had and a "nuc hive" is: The nuc hive ALREADY has a laying queen. So some brood is already developing on the bottom comb. Plus, I already have capped honey.

The nuc hive is an ESTABLISHED colony of bees (versus bees all together in a package with a queen they don't know and they HAVE TO ESTABLISH A HIVE). To put it in business terms: My previous hives were "start-up" businesses. This nucleus hive is an "established" business. So I'm "acquiring" "a going concern."

I am writing this outside on a stump where I can observe the bee gate. I opened it 15 minutes ago and want to make sure they come out and start bee-ing busy as bees.

They are circling, buzzing, taking out feces and learning the position of the bee gate for flight.

There are about 12 workers in the air above the hive that do nothing but circle and pirouette. I am sure they are fixing the location for the others. Some are spanning the three stories of our Victorian home further fixing the location.

In fact, I've read that the honeybee does many so-called "dances" like the one I'm watching now. I've already identified what I believe to be an IN-HIVE DANCE that I've called the Pollen Dance. Have, also, identified another in-hive dance I call a Collective or Group Dance. Some would maybe say the Pollen Dance tells other bees where the pollen source is. I think it is a happy dance and a dance that "tells" other bees "I've got pollen!" Like the elation of someone coming home with a big paycheck.

Professional observers, unlike me, have identified a Round Dance. The bee goes around and around in the hive in a rough circle. Perhaps she is communicating a foraging location. Maybe not.

Then there's the Wag-tail Dance (also considered a directional dance) where the bee makes a half-circle in one direction and then a half-circle in the opposite direction. These half-circles compose a whole circle. It's called "Wag-tail" because she makes wagging motions side to side with her abdomen as she's "dancing." Experts definitely believe this dance tells other bees WHERE the food source is. Perhaps. There are other dances: the Crescent and Pull dance.

I think what I'm watching is a flying Directional Dance. They are getting the lay of the land. They are learning where their home is. They are pathfinders.

An interesting observation:

When I first opened the hive, I noticed that Ed had put the brood comb on my observation side too close to the glass. Circa 50 workers are pressed against the wood at the top of the brood comb with their backs against the wood and are pushing with their legs against the glass. They are trying to MOVE the brood comb back about a quarter of an inch (!) because they love to nestle at the top of the brood comb at night. Often in the middle of the night I will open the hive and see hundreds of them huddled together in a long horizontal phalanx resting with their heads on the wood at the top of the comb and their bodies on the brood cells. I think that is one of my favorite sights in the hive. The little ones resting, touching.

For dinner we had those wonderful chickens they sell at the Yankee Peddler Fair. The men at the church BBQ them over huge charcoal pits covered with grills. They are juicy and herb-filled and pull-apart tender. Then we had generous slices of the famous Apple Pies the women of the church bake. They get together several days and nights in late winter and make and bake and bake. They freeze them and sell them. You have to order them weeks ahead of time or you won't get one. I always order three and take one to someone who needs a sweet and save two pies for my sweetie pies. Wish I had the recipe for the pie. The crust is soooo short. Of course, I could volunteer to make and bake them. You could do that, Sandra. Will have to have mother bake us a pie. She makes, hands down, the best crust I've ever tasted!! Her French mother, my grandmother, told her, "When you think you've got enough shortening, put in a little more." Mom's cherry pies---like home and flag--- (she doesn't use the canned goop but makes the filling herself with sour cherries from a can) are one of my two favorite sweets. The other one is the incomparable Turkish delight called baklava. I read that "they" think baklava was the first pastry. The reason given is that it is very primitive in its ingredients: phyllo pastry, nutmeats and honey. That means I LOVE the basics. I'm not an ice cream eater, but if I get it, give me basic vanilla. I can put hot fudge over it. When I was a teenager I would go to Isaly's and get a scoop of pineapple sherbet on the bottom and toasted almond fudge on the top. It had to be in that order. The chocolate top would melt down on the pineapple bottom and the tastes together were wonderful to me at that age. Now? Probably not.

June 14, 1983

Just got my first bee sting! Was walking barefoot over the clover in the yard and felt a little pick on my big toe. I retrieved the stinger, examined it under the high magnification and have put it in an envelope in this journal.

The stinger is the color of propolis and measures c. one sixteen of an inch long.

The end is indeed pointed, but the rest of it is spathe-like and, I think, has two horizontal lines in it. The part of her abdomen that is attached at the base and has come off with the stinger (thus causing her death) is clear colored and is an irregularly shaped mass.

When I was a little girl and would visit Meme summers in Zaleski, Ohio, I would occasionally get a bee sting in her yard or garden. She would always rub it in and say, "A bee sting is good for you, Sandy."

I know she didn't know why she said that and was passing on lore passed on to her, but she was right. There is something in the sting of the HONEYBEE, not other bees, that is life-prolonging. I read that there are more centenarians among beekeepers than any other occupation. Democritus (460-370 B.C.) was a Greek physician and philosopher. He always practiced a diet that included honey on a daily basis. He lived to be 110 years old. He probably got his fair share of stings.

There are, of course, a small percentage of people on the earth who are allergic to bee stings (from 1% to

3%). They can go into anaphylactic shock and die. One of Jesse's friends is allergic and when he comes over, his mother gives me a medication to give him should he receive a bee sting. However, the vast majority of people in the world are not allergic to bee stings. It would take over 500 bee stings in a really short period of time to kill someone. I read once that a man in Africa was stung 2,000 times and survived.

The Killer Bees, so called, are heading north. They are now in Central America and if not stopped, will soon arrive in the States. I'll have to research them and find out why they're called Killer Bees.

Note inserted later: Wow! They are aptly called Killer Bees. Since they accidentally escaped from a research facility near Sao Paulo, Brazil in 1957 they have killed 1,000 people! A biologist by the name of Warwick Kerr brought some African honeybees to Brazil in 1956 in order to inbreed them with the local European honeybees. He was trying to breed a strain of honeybees that would be more productive in tropical climates. 26 Tanzanian queen bees and lots of European honeybees flew the coop and now we're all in trouble. Those African queens have inbred in the wild with the local honeybees.

The African Honeybee (AHB) strain is more aggressive than the European Honeybee (EHB). That aggressive behavior means: they will attack without provocation and they attack en masse. Often if you inadvertently get near their hive, they will send out thousands of bees to sting you (sometimes to death). Run away? They have chased people a quarter of a mile! That's aggressive.

Since 1957 when they escaped in Brazil, the Killer Bees have progressed northward through South America, Central America and eastern Mexico. They travel about

100 to 200 miles a year! By the early 1990's they were detected in Texas, Arizona and parts of California.

But the big worry in the world about the Killer Bee is that because of its aggressive tendency, it will take over our native hives and threaten the US agricultural industry. It's amazing that over 15 BILLION dollars a year worth of fruit, vegetables, flowers and nuts in America are dependent upon beekeepers being paid to lug their hives to the fields so their honeybees can cross pollinate the crop. If and when the AHBs take over the hives in Texas, Arizona, California and Florida beekeepers will no longer WANT to keep hives with such aggressive bees in them. Without the hives, no oranges, broccoli, pecans. Got the picture. PLUS who wants to try to extract the honey from hives of Killer Bees. The honey industry itself is valued at 149 million dollars a year. The African bees don't store up as much honey as the European bees do anyway. So, pray tell, why did Dr. Warwick want to cross-breed them? We've got a problem, Houston.

June 22, 1983

There is a lot to tell that happened. I'll recount later. Right now I'm observing.

12:00 midnight. They have NOT been on the side that faces me. I think that is because Ed accidentally switched the bee gate and it now opens on the side that WAS in the kitchen.

Generations of bees favored the side that now faces the outside (which used to face me). They built on it and lived on it long before they used the side that faced out. I think those "vibrations," whatever, are still in the hive and the new nuc bees favor the same side.

So it has taken them some births and begettings to finally get around to my observation side.

55 bees on this observation side are building comb. They are not in their nice necklaces. That's serious, brood-comb business. Instead, they are one by one secreting wax from the little glands on their bodies and building the comb from the upper middle on down.

Great news re: PROPOLIS. I have become interested in the last months in this peculiar hive product and have read whatever I could find. It seems it is particularly known in Europe, Eastern Europe, for its curative powers. It is used, and successfully, for mouth problems, stomach problems and skin problems.

My sister, Tooie, was here and saw a spot on my upper right cheek. She said it could be a skin cancer. I went to Dr. Brooks and he said it was pre-cancerous and he would freeze it off or I could have it taken off surgically.

It was brown and Joel measured how many millimeters it was, etc. I told him I would like to try a cure I had read about---propolis. I explained to him what it was and he was amused. I asked him if I had time to experiment or if I should have it removed immediately. He said there was no rush. As I left, he said, "But do come back in 6-8 weeks and I'll take it off."

I got some more tincture of propolis from Ed. Every night I dabbed it on the brown spot before I went to bed. In the morning, I would remove the stain using alcohol and a cotton swab. I did this for five nights. After those five applications, the spot diminished greatly!! It was brown and raised slightly above the other skin. You could tactilely feel the difference in skin height and texture. Now the raised part has gone away completely. All that is left is a teeny brown freckle. I am really a fan of this propolis. Bees and bee products won't cure everything, but they will cure a lot of things. The beehive is the oldest pharmacy in the world.

Alexander the Great's body was coated with raw honey (in which were propolis and wax) for the long trip back to Greece from his death in India. Strabo, the great Greek historian (c. 64 B.C.-24 A.D.), says that he was placed in a golden coffin brimming with honey. Alexander had ordered before his death that his body be "embalmed" in honey. Aristotle, of course, was Alexander's tutor and had probably told him the "embalming" properties of honey. In fact, many ancient cultures coated the dead person's body with wax and then covered it with honey to preserve it. Honey's curative and preserving qualities have always been known.

One beekeeper I read about had a vat of hot water spill on him. He coated his arms, legs and stomach with honey and then proceeded to the doctor. There was hardly any damage from the burn.

I have found that the sticky juice of the household aloe vera plant is excellent for burns, also. Steve picked up a live log from the fireplace a couple of years ago. The flesh was raw and jagged. I kept coating the burns on his fingers and his palm with that clear aloe juice from the spiky leaves and in two days he had no marks!! For real.

When I went back to Joel Brooks about my pre-cancerous spot, he couldn't believe what had happened to the spot. He got out his notes and his drawing of the spot. He kept examining my right cheek---now smooth and spot free. Then, of course, he was not at all amused and was VERY interested in PROPOLIS. I had to explain to him what it was, what it was supposed to do, what exactly I had done and how I had done it. I'll bet someday doctors will begin to use propolis on skin cancers and mouth cancers.

Observation: Bees are not symbiotic like ants, but these bees have allowed a scared little fly that somehow got in the hive to stay among them for 7 days now. The fly has stayed on this observation side which is the least populous. The fact it has lived 7 days in the hive shows that it has been allowed to obtain food. The fly is skittish and scared, but the bees pay it no mind.

Since this side is blank, I can see through the comb to the pattern of brood on the other side. The brood is on the top of the brood comb and there are cells filled with pollen scattered throughout the middle and ends of the comb.

I am not feeding these bees any sugar water. They are totally on their own. I'm cutting out any chance for moisture!

Love that propolis smell coming from the hive.

I noticed some moisture in the lower left and right hand corners. I rested the wood panel on the outside rung so the air could get into the hive when the panel is over the glass. Now most of the moisture is gone.

This is a low-key hive. It doesn't as yet produce a frantic hum like the others did. They go in and out unobtrusively. I've never observed them luxuriating profusely in the morning sun on the window. Of course, there are only one and a half pounds of bees in there versus the three pounds of the others. That's probably a more comfortable number for this size hive.

June 24, 1983

Just saw the little fly. It's still alive. Who says bees aren't symbiotic. Symbiosis is when two dissimilar organisms live together. Ants are symbiotic. But bees are supposed to kill any insect or animal that comes into their hives. Even a honeybee from another hive that wanders in will be killed. But here they've let the bewildered fly live and sup with them.

Individually, one by one, they have made great strides in comb laying. They have drawn brood comb over the whole area except for the upper right and left hand sides.

June 28, 1983

Not one, but two small flies are in the hive with the bees! At least one of them has been there 13 days now.

It is a cold 64 degree, rainy day. They are hive-bound, working assiduously but silently. These bees are extremely quiet ones. I guess hives have different personalities like families do.

Re: the moisture. By not feeding them sugar water and not fully closing the wood panel on this side, there is no moisture on the glass!!

Just got a good close-up of a bee's antennae. It is composed of three parts: a shaft from its head, a ball, another shaft reaching to the outside world. The ball part seems to be the JOINT of the antenna, the point at which the two shafts pivot.

I have observed antennae going:

1. shaft, ball, outside shaft goes vertically down
2. shaft, ball, outside shaft goes to lateral left
3. shaft, ball, outside shaft goes to lateral right

I have NOT observed:
1. shaft, ball, outside shaft going vertically up
2. shaft, ball, outside shaft going horizontally back, i.e. parallel with 1st shaft

The bee's abdomen has:

Five black stripes counting her shiny black bottom. Usually, typically, the black bands go from light black near the thorax to darker and darker black culminating in the pointy, shiny black bottom. I have been able to distinguish different batches of bees from the last three stripes. Sometimes the stripes are not very dark and broad until the bottom one. Often the last two bands are black and so broad they almost touch one another.

Today's wax status: They have filled up most of brood comb except for very small parts now of the upper right and left hand corners. Fast workers!

July 4, 1983

Independence Day. Am having another pool party this afternoon. Kathy's Birthday is July 1^{sst} so I usually have a combination Birthday/4th July party every year. This year about a hundred people are coming. I rarely (once a year) have a party with just adults. I like parties that combine all age groups. Toddlers to be tended. Kids of all ages running around and in and out of the pool and dripping wet all over the Carriage

House and eating everything in sight. And, of course, the parents.

There are hundreds of bees on this side building brood comb. The honeycomb is still untouched, pristine foundation as when I put it in. They still build the brood comb starting in the center and working out. Centrists.

July 15, 1983

Today has been 95 degrees. It is 3:30 in the afternoon and a gentle rain is falling. I went to the bee gate. In 60 seconds I counted over 200 bees coming in out of the rain. But I counted 12 workers who were leaving the hive to try to get one more load before it gets worse!

July 19, 1983

In the garden I just observed a brutal and unusual example of the Fat Falling Before the Fittest.

A skinny wasp and a plump green cabbage worm were engaged in a life and death struggle inside one of the red cabbage leaves in my vegetable garden. The wasp stung the worm. The worm died. The wasp proceeded to saw the head of the worm off with its mandibles. As the sawing continued, a long (6") stream of dark red-green ooze leaked from the posterior of the light green worm. The wasp sawed off a half an inch chunk of the worm's upper body. The rest of the mangled form slipped into the bowels of the cabbage. The wasp flew away with its meal dangling from its mandibles. The drama was over.

A Cosmos in my Kitchen

We are promised in Scripture that the Lion will lie down with the Lamb. Maybe the Wasp will feast beside rather than on the Cabbage Worm, too.

Note: Bee Balm: I have a 7 ft. high stand of dark pink bee balm. My honeybees are rarely on it, but it is beloved of the hairy bees with bright yellow jackets. In Ohio when I was growing up, we called these bees, yellow jackets---appropriately, I think. Here in Connecticut they are called bumblebees. These loll around the bee balm and often are dead on it in the early mornings and will hang there for days until some other insect knocks them off.

I love the SMELL of this wonderful plant. The leaves and flowers are very fragrant and I rub that smell into my whole hands and smell it for about 20 minutes as I tour the gardens. The Indians made a tea from the leaves. Oswego Tea. I've made it---musty and earthy.

Note inserted later: The Connecticut Yankees are right and I am wrong! The furry big bees ARE called bumblebees and the skinny mean bees are called yellow jackets. They are both what are called Solitary bees versus Communal bees like my honeybees. All the solitary bees are indigenous to North America, but my honeybees are not. The Spanish brought over some honeybee hives in the 1500's and the English brought over their hives in the 1600's. By the time the pioneers headed west, there were millions of wild honeybee hives that preceded them. The American Indians called these producers of nature's sweet the "white man's flies." California, where lots of the pioneers were headed, could not be the leading producer of fruit and vegetables without those portable honeybee hives. Pollination and cross-pollination.

Liatris: My bees and all other bees LOVE this stunning, stately flower. One cold March day I broke two plants into 30 shoots and have got as many stalks of flowers---furry and pink, bloom starting from the bottom and going up, long-lasting, reliable, not temperamental, good cut flower---this Kansas Gayfeather. I think the pioneers called it that name. Imagine bumping around in a conestoga wagon and then rounding a bend and seeing millions and millions of these feathery flowers on such sturdy stems blanketing the hillsides!

I've always read that there are Pollen Gatherers in the hive and Nectar Gatherers. They are not supposed to gather BOTH at the same time. BUT right now there are honeybees on the liatris with packed pollen baskets. They are not getting pollen but have their nectar tongues out getting nectar. Are they only eating so they can continue gathering pollen OR are they getting nectar, too? Perhaps they can gather both pollen and nectar at the same time?

Nectar is composed or 60-80% water and the rest are sugars (sucrose, fructose, glucose and other sugars). Just think: nectar is mostly water with a little bit of sugar thrown in. Just like maple syrup. The sap from the maple tree tastes like water really with a teeny hint of sweet. We boil it and boil it and boil away all the water and voila the wonderful maple syrup. It's one of my favorite tastes. The honeybee is, of course, the only way that the nectars in flowers and plants around us can be collected and deposited and capped and---honey! On his own, man can't harvest nectar. But the Europeans learned from the American Indians how to harvest the sap of the maple tree and get its sweet. The Indians used to slash open the maples trees. Most of them died. The white man improved on the technique by tapping into the tree with a hollow wooden peg and getting out the sweet sap coursing through the cambium layer. With that innovation, the same trees could be tapped year after year.

Other flowers on my property beloved of my bees: ilex, heaths and heathers, lavender, the oregano bloom (pretty pink clusters), clover, flowering kale, wild asters, the superb bloom of the leek (which I use as a cut flower, dries beautifully), chives, allium senensis, white hydrangea which is right outside the bee gate and a favorite for its pure white pollen, cosmos, mint blooms, squash blooms, cucumber blooms, tomato blooms. But there is no question that the smaller the flower the better my bees love it. Their beloved liatris is made up of 100's of little flowers.

On 6/22 I entered the observation about the fly that they have allowed to live with them. He is STILL there----healthy, getting bolder, climbing all over the glass. That's a month of cohabitation now.

August 2, 1983

This nuc hive will never be full enough to swarm, but 5,000 is a good number for the three brood combs, three honeycombs that I have in my observation hive.

I've, so far, solved the moisture problem:
1. No sugar water feeding which eliminates possibility of dripping.
2. Cocking the wood panel against the glass rather than fitting it over the glass. Air can move in and it is not airtight.
3. I do think the limited number of bees has a lot to do with it, too.

Still no comb laid on this side of honeycomb. There is now a thin layer of wax over all of the brood comb with more build up in the center.

They have brought in propolis, nature's own antibiotic. There is a thin layer on the left side of the outside of the honeycomb and thin layers on relevant points of the brood comb. They, also, put propolis at the entrance to the bee gate. I think this is because when they RETURN from foraging, they MAY have picked up some germs. Passing their legs over the propolis as they enter would probably kill those germs. "Propolis" in Greek means "before the city."

Propolis and its amazing properties was known to Aristotle (384-322 B.C.). He describes it and proscribes it for healing in his Historia Animalium. The ancients used propolis primarily for healing wounds and sores inside the body and outside the body. This bee glue was used frequently in bandages and poultices. Since it is a natural versus a synthetic antibiotic, you cannot build up immunity to it. Plus, propolis seems to kill only the bad bacteria. Most antibiotics will kill all bacteria. (We need "good" bacteria in our body.) And unlike so many drugs, propolis boosts the immune system rather than depressing it. The principal ingredients in propolis that boost the immune system are flavonoids. Flavonoids are antioxidants and help to trap free radicals and bind bad metals so they don't hurt your system. Flavonoids were only identified in the 1950's. The ANTIOXIDANT property is not what interests me, however. It is and has been the ANTIBIOTIC properties of this resin---its potential for killing bacteria and germs. In the Boer War (1880's), for instance, they mixed propolis with Vaseline and applied the mixture to wounds as an efficient disinfectant. I read that it is the organic acids and terpenoid compounds that create this antibiotic quality in the propolis. Those I know nothing about. But I do know that propolis needs to be discovered!

When I open the hive, there are c. 30 bees on this side. Within 30 seconds, a hundred have arrived and within 5 minutes, hundreds come. They hurriedly enter as if checking, guarding. They climb quickly over the glass and observe me observing them. They see there is no danger and then begin to work. By opening the hive, it seems I call their attention to this neglected side.

Today I saw a gorgeous jet-black butterfly with bright yellow markings on her wings and body. It was resting poised on the echinacea (another of my favorite flowers with its gleaming, prickly rust cone and pale pink petals). I show Jesse how he can comb his hair with the cone. The butterfly had 6 legs, 2 upright feelers and was probing the heart of the cone with what looked like a feeler from the center of her head. I don't know anything about lepidoptery, but it was beautiful. Made me wonder what they use to get nectar and pollen. Was that prober a proboscis? Made me want to collect butterflies, but I couldn't kill them and pin them down. I do think it's okay to do that, but not for me.

In the batches of bees I've had there are creative differences in modus operandi. These bees, for example, not only cut away comb on the sides for entrances and egresses, but they have chewed all the way through three hexagons on top of the brood comb and use these eight-sided doors to squeeze from one side of the comb to the other. My other batches never did that.

It is so wonderful to see the hive clean and dry as a bone!

August 5, 1983

Today I saw another butterfly. I watched it feed on a zinnia. It does appear to have a proboscis. The butterfly, unlike the bee, is very delicate. It seems precariously put together whereas the honeybee is a sturdy little bullet. The butterfly's prober (proboscis?) goes down into the flower almost totally. Is that part hollow? I don't observe the butterfly taking the tip and putting it to her mouth after probing. Have to get a book on butterflies someday.

September 9, 1983

I simply don't believe it! The fly is still alive in there. It is perky and very integrated into hive life. It goes all over the glass and combs and goes in and out the passageways. No---TWO of them are there!! Just saw the 2nd one again. That's two small flies living in a beehive since June 13th---almost 3 months. They may not give the bees anything, but the bees have provided for them so that is somewhat symbiotic.

September 17, 1983

4:15 P.M. Unusual: First of all, I saw a bee frantically carrying a larva around the comb. The larva was wormy, white with two black eyes. She carried the larva up and down the comb in her mandibles until all except a small piece of the larva broke off and fell to the bottom of the hive. She continued and still continues to parade that shiny white piece up and down the glass and over the comb---frantically. Is something wrong with the larva? Did something harm the larva? What's going on??

Secondly, I see another type of bee---very painted, bright yellow and black about the same size as my bees (a yellow jacket?). It is 65 degrees and cloudy, misty, not good foraging weather. Maybe that is how the other type of bee got in---any port in a storm idea.

The intruder/confused bee was there c. 2 minutes going back and forth seeming to look for an exit from the hive. There were c. 25 bees on this side. They continued working unfazed by the Larva Parade and intruder.

Suddenly, about 8 bees started tracking the intruder. They got her down to the bottom of the hive and surrounded her. I can't tell how many stung her. I do know for sure that one did because she died. One of my other bees picked up the honeybee that had stung the intruder and carried the dead heroine around the hive parading her body. I've seen them do this a lot with their dead. It's as though they are letting all see the dead. I remember hearing about Roosevelt's train carrying his body from the summer white house in Warm Springs, Georgia where he died back to Washington. It stopped at all the little stations along the way. Perhaps this is a similar type of ritual, this Parading of the Dead. A traveling homage to heroes/heroines.

I remember the day Roosevelt died. I came home from school and my mother was in the sunroom crying. I ran to her, "What's wrong, Mommy?!" She looked up and patted my face, "Oh, Sandy, President Roosevelt has died." I was so relieved. I thought something really bad had happened. Then I wondered why my mother was crying because we weren't Democrats. My mother had turned against Roosevelt when he tried to "pack the courts" whatever that meant. Looking back on it, it was wonderful that my mother was touched by the death of a man she did not approve of. Now our politics are so balkanized. And we know every intimate detail about our Presidents. It would be very hard to hide another Lucy Rutherford from the American people. The press hid all Kennedy's women for about 10 years and now these women are all coming out of the woodwork!

Anyway, back to the Parades. When the intruder was stung, she became weak. She, in turn, was picked up and carried up and down the brood comb by a bee. After that Parade, 4-5 bees surrounded her and pinned her on her back. As the intruder was dying, the dying worker who had stung her was still being paraded. What drama!

This all lasted about 5 minutes. Then both bodies were disposed of. The bee that sounded the alarm by carrying the larva like a flag is gone. The intruder, the heroine and the heralder of trouble have all gone.

I'll bet the intruder tore the larva out of the brood comb. The other bee carried it all over the hive to show that there was an intruder and this is what she had done. The guard bees showed up, tracked the villain, killed her, paraded her dead body AND the dead body of the heroine who had killed the intruder. Entrance, Abortion, Alarm, Attack, Self-Sacrifice, Deaths, Parades---all sounds pretty human to me. It was fascinating to watch!!

I was thinking that the greatest intruder to hives is man. For thousands of years he has raided hives in order to satisfy his innate "sweet tooth." Near Valencia, Spain a cave painting was found in 1919. Some have dated the primitive red ochre painting as far back as 15,000 years ago. I'm looking at the painting now. It's called the Bicorp Man named for the little town near the cave. The very crude painting depicts a naked man with long hair. He has climbed three sturdy strands of rope and is very high up in the air. He is gripping the ropes with his knees. His left hand holds a basket with a handle on it. His right hand is reaching into a hole in a rock or tree. Many marks are swirling in circles around his right hand. Giant marks are at his back and legs. He is clearly raiding a hive for the honeycomb. The bees are stinging his right hand and are threatening to sting the rest of his body. Most prehistoric cave paintings at Lascaux in France and at Altimira in Spain show ONLY animals and human handprints. People believe that these paintings represent the animals they hunted and commemorate the bravery of the hunters. I don't think, however, there is another cave painting in Europe that depicts a man actually DOING a heroic act. That the raiding of a honeybee hive is commemorated at all shows that those societies thought the act was at least as brave as killing a reindeer or aurochs (extinct type of cattle). I think of the day the Bicorp Man decided to climb those ropes. Someone had seen the honeybees way up there. Who would get the sweet? "I'll go," he said. And they had to rig the ropes and get the basket that would carry the honeycomb down. His whole family or tribe was watching at the bottom of the ropes as he climbed the ropes and steadied himself outside the hive. He pushed his right hand into the hive. They heard the yells as he was stung. They admired him as he reached in over and over again to get every bit of honeycomb. And he slid down the ropes quickly fighting off the bees every inch of the way. His friends and family ran for safety. He reached the ground and ran to his cave. Some bees were still clinging to and flying around their honey. People shooed them away. Stomped on them. Smacked at them. And then---they picked up a piece of the honeycomb. YUM! As a woman tended the Bicorp Man's stings (probably with mud poultices), everyone congratulated him and dipped their hands into that gooey, sticky, golden treat. They saved, of course, the best strip of comb for him. And they painted what he did on the walls of the dark cave nearby for all to see and to remember and to admire. I love that brave man!

September 24, 1983

One fly is still alive and totally at peace in the hive. It was crawling along the bottom of the hive as nonchalantly as you please, brushing against the bees, accepted by all.

September 26, 1983

I still can't get the little Hive Drama out of my mind.

The larva was carried like a banner as was the intruder and the one who gave her life for the hive. This frantic parading about in the mandibles was a "showing" of what was going on to all the others in the hive.

I have often observed them pick up a dying sister (never a drone) and parade her up and down the comb and glass. Once the bee is shown and dead, that's that and she is carried out, flown away or dropped to the hive floor.

The yellow jacket intruder obviously scavenged the larva out of its cell. He was then driven away and dropped the larva. The larva was picked up and shown to all. The intruder was cornered by 8 sisters, stung, pinned down until almost dead. Dying intruder was paraded around the comb, died and was disposed of.

The amazing thing to me ALWAYS is that only the bees APPOINTED for the task take part. It's not a free-for-all where everyone tries to get the intruder. The other bees always continue their work as if nothing were happening. During swarm even, the ones appointed to stay just continue cleaning brood combs and meandering around as the others are literally going crazy.

Now in this way, they are totally UNLIKE the human community. If there is human drama, everyone in the immediate vicinity responds some way.

Yesterday I went over to Rick's and watched him get into his two outdoor hives. Carole originally got them, but she was stung a lot, so Rick took them over. He put a rope in his smoker to burn. It seems to work very well. I had him give me a big ball of propolis. He didn't know what it was, where it came from or how efficacious it is. Like so many who keep outdoor hives, they don't KNOW the bees, but they do KNOW more than I do about manipulating bees, that's for sure. I have no desire to do that. I was very impressed with Rick's cool expertise as he worked.

September 28, 1983

Since June 11, my bees have literally been on their own. I have not given them any sugar water. The hive is totally dry and they have been healthy and moisture-free.

Today I'm going to introduce a half a bottle of sugar water and see if it makes any difference. (Ed and all the books tell me to keep giving them sugar water.) I would love them to lay comb on this side, so I could SEE them all the time. Maybe this will give them the impetus, the fuel they need?

On the empty, untouched honeycombs I notice they have deposited in the last 24 hrs. four loads of those small, white granules which I have seen in my other hives. They don't look like any pollen I've seen and I don't think they are wax. They are granular rather than powdery like pollen. Just noticed that they have got one repository of granules in the center of the brood comb, too.

5:30 P.M. I have had the sugar water in for 10 minutes now. There have been 4 bubbles that have floated gently to the top. But I have noticed that NO bees have touched the nozzle. I spy a small puddle on the floor of the hive. I'll leave it in for a few hrs. and see what happens.

7:00 P.M. The sugar water has created quite a stir now. A different type of Intruder!? There are 8 bees around it excitedly taking turns sipping. The puddle of sugar water at the bottom of the hive is gone. They have absorbed it with their nectar tongues. The bottle has gone down one eighth of an inch in one and a half hours.

As of now, I've observed that only 2 bees can sip at a time from the nipple of the bottle.

I've never seen this batch of bees so excited, this active. It is dark outside. But there are c. 10 guard bees out on the window excitedly running up and down. Is the sugar water really an Intruder? And are they trying to SEE from the window how it is getting in?

Whatever the facts are, the introduction of sugar water is an event for them. Since so many are out crawling on the window, I wonder if there is a smell memory that the others have left on the nipple saying, "Beware!" I don't know if they are happy for supplemental food or if they "know" this little bottle's drippings helped destroy bees? Does it?

A Cosmos in my Kitchen

I'll probably never know.

11:30 at night. The introduction of sugar water into hive continues to breed interesting behavior. It's late at night and 50 degrees outside, yet there are guard bees on the outside of the hive on the window in a military formation. Maybe this will be an all-night vigil. They've never done this outside, nighttime vigil before.

Since the 5:30 introduction of sugar water, they have consumed three quarters of the bottle. At 7:00 there were 8 bees sipping. Now 3-4 bees are working the nipple full-time! They seem bent on emptying this intruder as fast as possible. Maybe they're wise.

Two of them put their nectar tongues in simultaneously. They fill up and then another two work the nozzle. Most stand on the right or left of the nipple and sip. Some lie on their sides like a baby nursing a bottle. Others crawl on top of the nozzle and put their nectar tongues in from an upside down position.

I've just got a flashlight to work and can now see very well. The floor of the hive is dry but messy: flecks of white wax, globs of brown. It is most messy on the nozzle side (the right) where they have been consuming. The mess is not serious, however.

I'm kneeling on the floor observing and have noticed debris on the windowsill under the hive. When I cleaned it up, the debris consisted of wings, legs, bits of same white and brown debris on the hive floor. NOW HOW IN THE HIVE DID THEY GET THESE THINGS OUT OF THE HIVE AND ONTO MY SILL INSIDE MY KITCHEN? The hive is pretty airtight and there are not a lot of joints in it. But they have found a way to shove these bits out. Maybe through the bee gate slits? Maybe between the window and the board upon which the hive rests?

Why don't they just take them outside? Maybe it is too cold.

I've noticed repeatedly in the last year and a half that they manage to get propolis through or under the glass (has to be under) and onto the bottom of the recess that the panel board (which covers glass) rests on. I've not mentioned it before (haven't mentioned so much), but have been flabbergasted at this seemingly miraculous feat. The glass fits VERY snugly in its groove. You couldn't put a hair under it. Yet they manage to shove a good deal of propolis out of the hive and into the outside groove.

Also, several times propolis has been on the observation glass on MY side of the hive!! I've scraped it off with my fingernail and tasted it---it was propolis all right. There has to be some sensible answer to the question: How could propolis get out of the hive and into the kitchen? Period. How can these things get shoved out of the hive? I couldn't shove a bee's leg back into the hive. How can they shove a leg or wing out of the hive?

Maybe propolis can be in a gaseous state. It escapes the hive as a gas and then condenses and hardens into substance outside the hive on the glass? That's the only logical explanation I can think of on a chilly Fall midnight.

It's past midnight. The guards are as they have been on the window---immobile, rooted to their posts.

Sometimes I get discouraged when I open the hive and see not one bee on all these blank, undeveloped cells. But then I've gleaned so much from the little

I can observe of them that I'm thankful for what I have!

I know that my bees know me. Bee people are sure that the bees know their keeper. I read one account about a woman who kept bees and died. The man who took over her bees put his scent on a twig of rosemary and went with it to the hive and "told" the bees that she was dead and he was their new master.

Wonder if that worked? Also, they recognize each other by pheromones so why not recognize the keeper by her smell. As a human, especially when I was young, I could tell my mother's smell and my father's smell. When my children were young, I knew each of their smells. I can still faintly "smell" Kathy and Blake even though they are older. Jesse, of course, I can smell.

When I come upon a honeybee in the garden, I often talk to her and will "pet" her. I pet her on the back with my index finger. Some have moved away as I do this, but most allow me. They are very gentle with me as I am with them.

I noticed today they were gathering nectar on the dahlias, zinnias and the flowers of the basil herb. They are getting pollen and nectar from the yellow spoon mums. They love the nectar on all alliums and were on the last blooms of the chives today.

Nectar, nectar. Keep bringing in that nectar. It's one of the life bloods of the hive. An average hive of honeybees (my observation hive is not average) brings in about 528 lbs. of nectar a year!! If they can only carry several ounces of nectar each time they go and return, think how many millions of miles in the aggregate they have to fly to get 528 lbs.! And the 528 lbs. of nectar is mostly water. When that amount is converted to honey, that's 132 lbs. of honey a year made by the normal hive from 528 lbs. of nectar gathered! Don't forget those women are out there gathering pollen and propolis, too. It all puts most human women to shame.

I think I know what the white granules in the hive are---hydrangea pollen. The 100 yr. old hydrangea tree is blooming. Between our property and the West Lane Inn next door there must be 60 hydrangea trees. They are storing up on that big white seed of nature.

I just shined the flashlight on the outside window. Up flew a guard bee from the entrance and beat against the window. "Stay away," she says.

And so I will. Goodnight, my sweets.

September 29, 1983

It is 11:45 in the morning and they are still sipping away on the sugar water.

I mentioned the guard bees on the outside window last night. Today it is 62 degrees and sunny. About 8:00 this morning I noticed they were still on the window---immobile. I came home (c. 11:20) from taking Jesse to school and then going to the coffee shop for coffee and newspaper (my routine). I examined them with the magnification and found no sign of life, but I felt they were still alive. Maybe they were in a type of suspended animation (?) from last night's 40 degrees temperature? How could I give them the heat they needed to limber them up, to unfreeze their bodies? My hands.

So I started the process of putting my hands directly against the window right where each body was. I did this for c. 2 minutes and found one moved a leg.

I started applying my palms to all the areas of the window where they were and, also, to the surrounding glass. The end of the matter is: after 12 minutes, guards 1, 3, 4 & 5 all revived and flew down into the hive. No. 2 is on the metal frame of the window where I can't reach with my warmth. She is still immobile. They were all there at least 16 hrs. (c. 7:00P.M.-11:00 A.M.) without food. I'm sure they are thankful to be back home. They were willing to give their little lives to guard the hive. They went out, positioned themselves, got "stuck" in the cold weather and would have died there if I (or the sun) hadn't revived them.

The sun was not strong enough on the window to do the trick. None of the other bees had come to their assistance, though. They could have come out and given the needed warmth, but didn't. Wonder why? Guarding is always hazardous, lonely duty as can be seen.

As I read over the above notes, I see the numbers 1, 2, 3, 4, & 5. Again I'm reminded of Fibonacci and his numbers. What fascinated me about the Fibonacci number sequence was that he was able to find a mathematical tendency (rather than law) buried in the petals, sections, etc. of the plant world. But Fibonacci, also, has a lot to do with 0, 1, 2, 3, 4, 5, 6, 7, 8, 9--- our Arabic numerals. When he was a child, his father worked in Algeria. Fibonacci was enrolled in a math school there in Bougie. The Arab number system (or "Hindu system" as it was then and still is called by the Arabs) was not the Roman numeral system he had learned in Europe. As a young math prodigy, Fibonacci immediately saw its superiority to the cumbersome Roman system of calculating. 2 (one stoke) in the Arabic system was II (two strokes) in the Roman system. The simple symbol 8 (one stroke) in the Arabic system was VIII (4 strokes) in the Roman system. The Arabic system took much less time to calculate and was more practical. Less lines of writing=more efficiency. It was better for commercial ordering and filing and billing, easier to weigh and measure, to calculate interest, etc. He traveled throughout the Arab world learning the intricacies of the new system. When he was 32 in 1202, he published his historic (for the history of mathematics in the West) book <u>Liber Abaci</u> (in English <u>Book of Calculation</u>). European mathematicians immediately embraced the new Arabic system of numbers. With those sparse ten symbols, they could calculate numbers to infinity. The general public did not use them for another 300 years. ("General public." That would have been me!) Western mathematicians loved the simple system. For instance, Fibonacci published his book at age 32 in Arabic numbers; XXXII in Roman numerals. The book was published in 1202 (Arabic numbers). That would be in MCCII (Roman numerals). When I look over the numbers on these pages, I'm so thankful that Fibonacci introduced us to the Arabic system (that they learned from the Indians that they learned from the Brahmi). I know the Roman numeral system (took 3 years of Latin), but when I think about doing the simple math I do in my daily life with Roman numerals versus the Arab numerals, I'm boggled by the simplicity of the Arab system!!

<u>ARABIC #'S</u>   <u>ROMAN #'S</u>

Have: $3,000. $MMM.

    -   $734.   - $DCCXXXIV.

Equals $2,266.   Equals $MMCCLXVI.

September 30, 1983 (September XXX, MCMLXXXIII)

As I continue this Arabic numeral vs. Roman numeral fun exercise, I can think of several places where we still use the Roman numeral system. When I was younger, the date of publication of a book was in Roman numerals. 1955 was MCMLV. Some watches and clocks have Roman numerals on them. 12:00 is XII. Pope Pius IV, Elizabeth II, John Jacob Astor III. Important 2nd, 3rd, 4th and 5th same-name people have Roman numerals after their names. When I do an outline, I use Roman numerals: Part I; Part II; Part IV, etc. When we use Roman numerals, we are really making LETTERS in the alphabet STAND FOR NUMBERS. When we use the Arabic numerals, SCRIBBLES STAND FOR NUMBERS. I like scribbles best.

Have forgotten to mention one of the most important by-products of my bees: apples. We have three (III) old apple trees, so old they are hollow on the inside. Rotted chunks of them fall off every year. We've been here 7 (VII) years and even though they bloom their gnarly branches off each year, not a single apple have we ever received. Reminds me of one of Jesus' dicta about how to tell good from bad: "..know a tree by its fruit." Matthew 12:33 (XII:XXXIII) Which means it takes time to determine if one lives by what one preaches or if one just preaches. Have seen a lot of beautiful blooms in my life, not followed by any fruit.

So were my old apple trees. Well, this year they are loaded with apples. Unsprayed, wormy and a bit pocked, but still good eatin' apples. In the spring, the bees loved the blossoms. I did, too, and even candied them along with violets, johnny jump ups and some forget-me-nots. Now I see in my own backyard the fruit of cross-pollination!

At the Big E Fair in West Springfield, Massachusetts (II s's and II t's) this year, I read that bees through cross-pollination are responsible for c. 70% (LXX%) of the crops in America!!

Workman Publishers said today they "may" be interested in a book about bees I've been working on: <u>The Heart Of The Honeybee: An Illuminated Ms.</u> It is informative and CRUDELY illustrated by me. Jesus, You'll have to help me if You ever want any of the many things I constantly write to be published because I don't have the DESIRE to do what needs to be done to hustle a ms. I just have the DESIRE to write!!!! Thank You for that desire and the joy it has given me in my life.

I'm just finishing up this second Bee Book of Observations. In the back I've catalogued some times when I've tried Propolis as a medication:

1. With a pre-cancerous growth on my cheek. June, 1983

    Applied 5 nights before bed. Removed propolis with alcohol in the mornings.

    RESULT: It went away after five applications. The raised part went away and left a faint brown spot which completely disappeared in 30 days.

2. Kathy had a bad case of cramps. I put three drops of propolis on a cookie and she ate it. July, 1983.

    RESULT: Cramps went away in two hours vs. her normal all day cramps.

3. Jane T. had the flu. Hadn't eaten in several days. I put three drops on a piece of bread. She ate it. July, 1983

RESULT: Reports her stomach was okay in 4 hrs. and she began to eat.

4. Lena had a psoriasis-like rash on her arm and side for two months. Applied propolis with alcohol directly to rash five times in a row.

   RESULT: Rash was healed. Red spots remained. Red spots went away in two weeks.

5. Marv S. had a brown spot on his face. Gave him propolis and told him to do what I did. July, 1983

   RESULT: He didn't do it. Spot remains.

6. I gave a bottle of propolis to Natalie C. who was riddled with cancer. I told her to take three drops a day on a bit of bread. What had she to lose?

July, 1983

   RESULT ENTERED LATER: As of February, 1991 Natalie is alive and well. The cancer went into remission and has not reappeared. She told me she took the whole bottle, She did have conventional therapies. Perhaps the propolis aided them?

October 10, 1983

This new bee book is Bee Book Three, III, B3 (as in Vitamin B---because I've learned so much about the vitamins, minerals, acids, etc. in honey and pollen and propolis). It is a normal-sized book with cream cloth covering with flowers tumbling down. A nice feel to it. Hope there are nice things to report in it.

Here's a list of the contents of JUST ONE SPOONFUL OF HONEY: water, 7 kinds of sugars including dextrose, levulose and sucrose; 12 kinds of acids including amino, citric, formic and acetic; numerous minerals including calcium, sodium phosphorus, potassium, sulfur, magnesium, iron, copper, manganese, silica, silicon, chloride; all kinds of proteins; vitamins including thiamin, riboflavin, ascorbic acid, pentothenic acid, Vitamin K and biotin; enzymes including phospatase, diastase, invertase and catalase, terpenes; aldehydes; acetylcholine; some propolis; some pollen and some wax both of which contain more vitamins, minerals, acids, enzymes. Wow! I'm going to take a spoonful of raw honey every day. Maybe? I'm not good at resolutions like that.

Just gave them another half bottle of sugar water. It takes them c. 25 minutes to notice its presence. When I opened the hive and inserted the bottle, there were 2 bees on this side cleaning half-clean cells. The bottom of the hive was clear of bees as were the sides. Very lonely and inactive. Now there are 7-8 scurrying workers on my side. They are very excited. There are 3-4 bees nursing the bottle. There is activity. One of the bees who was cleaning continues her task, unperturbed. I know the excited ones are the guard bees. I can see them coming up from the bee gate.

The guard bees have resumed their positions as of the other day on the outside of the hive on the window. The introduction of sugar water really calls out the troops! There must be 12-14 outside flying around. I still can't figure out if they are pleased or threatened. Both could elicit that response, but the military formation of the guard bees shows me it is maybe a threat.

Before I inserted the bottle, I upended it and let the sugar water drop out on my hand until a vacuum was created. When I put it in, only one drop went on the hive floor.

Now inside the hive and outside on the window is buzzing. I love that sound! Ten are now nursing the bottle.

They've stored a lot of white hydrangea pollen on this side. There is a thin coat of propolis on all the untouched honeycombs, around the rims and some on the inside of the cells.

Yesterday I had the nicest experience. I was near the yellow mums and saw them gathering nectar and pollen from them. So were some little ants and other insects. There's not a lot of bloom left. But they can still gather on the zinnias, cosmos and weed asters.

Anyway, I was buzzed by a bee and ducked. She came back. I put out my hand and "she came to me." She landed, crawled on my index finger and then on my palm! I talked to her and then she flew away. I petted several other bees on the spoon mums. I went to the waning vegetable garden. Another bee (or was it the same one?) buzzed me. I instinctively ducked. She came back. I held out my arm and she came to my hand, crawled around on my palm and then spread her wings.

"Active, eager, airy thing
Ever hovering on the wing."   Aristophanes (c.446--c.388 B.C.)

They've let me pet them a lot and I hang around and among them, but they have never come to me in that way---getting my attention, so to speak, and then landing on me, examining me and flying away. I wonder if it is because they "know" I resuscitated those guard bees on the window? Or was it one of the guard bees? Or just something else? The guard bees on the window today should do pretty well because it is in the 80's today. A New England Indian Summer day.

They have posted guards in exactly the same positions that they did the other day. Maybe this is a good military defensive formation, too? I know in WWII the bombers flew in a V formation. They were emulating geese flying. There is a lead plane and pairs of planes fan out on either side of the lead plane to form the V. Of course, it is a logical formation. Each plane (goose) can keep in visual contact with every other plane. Except, of course, the lead pilot.

Langstroth's book has a picture of a guard bee in defensive position. She is firmly planted on her back two legs. Her two front legs are lifted up (ready for use) and her antennae are pointing aggressively toward the intruder. She's a formidable looking one.

One old worker is dying. I've not seen anything like her before. Her abdomen color is bleached out. The stripes are gone. She's grayish tan. Her wings are a sight---tattered and torn to fragments! She feebly walks up the right side of the hive licking the propolis. I feel even that balm will not heal her. Rarely have I seen such wear and tear on a little bee's body!! She must have braved much and often for her people. A right wing has holes and much of it is missing. A left wing is half the size it should be. The rest has been torn off. God bless her. She has 4 wings and has been flying or fanning for all of her 6 week life. Her wings whirr at up to 11,400 times a minute. And think of her flying through the air at 12 mph day after day battling winds up to 25 mph! She deposits her load of nectar and pollen in the hive and goes right back out. Teeny as she is, she compulsively takes on the world! And she DOES look the worse for wear as so many of us do after a long life of service!

I have a bit of propolis in my mouth. Sticks to my teeth. Nice spicy bite. Has really helped my sore throat and swollen glands. In bee circles and such propolis is called "Russian Penicillin." The Russians have done extensive studies of propolis and its effectiveness.

It does have antibacterial properties. I much prefer something from nature to something manufactured when I'm sick or ailing. But I'm not crunchy. Not a Naturopath. Just think it's more LOGICAL. I read once that the cure for something you get in nature (e.g. hay fever) grows right beside what gave you the sickness. Goldenrod and ragweed can cause sneezing and hay fever. Growing around those plants are the cures. Who knows?

A small ant has found her way inside the hive. She's on the floor. She's crawling on the glass. They're aware of her but are leaving her alone.

October 4, 1983

They drank the half bottle of sugar water in 8 hours. They worked full-time with 6-8 nursing the bottle.

The guard bees were still immobile. They were rooted to their posts at 8:00 this morning. By 11:00, they were gone. It is 80 degrees and sunny, so they didn't need my help defrosting.

I've taken the bottle out and put in the clay plug. About 6 of them are picking away at the clay.

There are two dead on the bottom of the hive, dismembered with the wings and legs removed. I've found some more wings and legs under the sill on the inside of the kitchen. Still don't know how they get them OUT of the hive and INTO my kitchen! (?)

A bee just came in laden with yellow pollen. She went to where the bottle was, examined the area and left.

The little ant that got in late last night is still in there wandering around. They have allowed her to coexist.

That's two flies and an ant coexisting with them. I'll see how long she stays. She's so small that she could crawl through an air vent and get out if she wanted to.

NO---there are two ants in there! Just saw the other one on the bottom of the comb. More than two! On examination I've seen at least 6 now. I started looking around and found 6 dead ants under the hive in my kitchen where they dump legs, etc. They probably killed them and then deep-sixed them outside the hive into the kitchen. I'll look tomorrow and see if there are more. There probably shouldn't be ants and flies in a hive. Portent!?

When they saw my flashlight on the sugar water hole, about 6 of them appeared and started crawling around the hole. Expectantly or...?

Note: I just read the most incredible (meaning "not believable") thing about ants. The total weight of ants in the world is EQUAL TO the total weight of human beings in the world!! Is that true? I have no way of checking this. But I do know there are a lot of ants in the world! They are indigenous everywhere except in very cold places like Antarctica and, believe it or not, Hawaii! Unlike my bees, their colonies are often composed of millions of ants. They do have a queen, female worker ants and males like my bees. They are in the same Order as bees: Hymenoptera. Their family is Formicidae. I've eaten a few black ants in my time (experimentally). There is a slightly acid taste. Ants produce this formic acid and it is used occasionally to preserve animal feed. But the thing that has always fascinated me about ants is how they "farm" aphids. Aphids are those nasty looking teeny insects that can pock your flower garden and kill your vegetable yield. I once saw a picture INSIDE an ant colony. The aphids were hanging from the ceiling and the ants were

MILKING them to extract honeydew. Honeydew is the sweet liquid that exudes from certain plants in the summer. The ants love honeydew and have enslaved aphids to get the honeydew. The ants take the aphids out of their colony in the morning and put them on certain plants. (!) The aphids milk the honeydew from the plants. The ants milk out the honeydew from the aphids. (The Great Chain Of Being.) That's credible. Believable. What do the aphids get out of this symbiotic relationship? Protection. Ants protect feeding aphids from other insects that prey on them. The ants drive them away or sting them to death. Slavemaker Ants steal brood from other ant colonies and raise the brood as their own slaves! A type of ant spends its entire life riding on the backs of other ants. There must be a benefit to the ant carrying the other ant, but I can't imagine what it could be. Just like I don't know why those ants are in my beehive and why my bees tolerate them. So many worlds! So much to learn and study. So little time.

October 5, 1983

Gave them half a bottle of sugar water. They noticed it immediately. Like Pavlov's dog, they are now conditioned to where it is, will be and what it is. I put it in this morning to see if they would set guard bees out like they did the other two evenings.

10:40 A.M. There are at least 20 bees hovering around the nipple. The hive is much busier with the ready source of "nectar." I think in the daytime many more are available to nurse it. We'll see how long it takes to finish this one. In the daytime they have posted no guards.

Last night in preparing my illuminated Bee Book Ms. to submit to Workman, I read through Bee Books I and II. What a lot of information. What excitement, drama, defeat and triumph. My bees have given me a COSMOS IN MY KITCHEN---good title for a book.

Today I noticed what I have observed over and over but maybe have not mentioned:

Pollen carriers, e.g., will go from flower to flower getting Nectar not Pollen. I feel this is for their own energy as they work---sort of like a lunch break. I don't think they're taking it back to the hive. I don't know if nectar gatherers also get pollen. It's harder to tell the nectar carriers except for empty pollen baskets. But pollen carriers do get nectar for whatever reason.

When a worker dies, her proboscis extends. Normally, the proboscis is folded under her mouth and onto the thorax. When she needs it, she just flicks it up and it extends. But at death, it lets go, extends outward like a sword from her mouth. Must mean that it takes muscular energy to keep it folded and protected.

I am now observing as I have for a year and a half the Parading of the Dead Ones. This dead bee is being held by the live bee's legs. Usually, the dead one is held in the bee's mouth. Just think, if we did this, I would have to pick up in my arms or by my mouth a dead person who weighed 110 lbs. and parade her all over town. A very weighty job!

October 15, 1983

Gave them sugar water. (Ed said so.) Took them 30 minutes to come to it, but after one minute, a guard noticed it and she marched around the bottom of the hive. They've posted 6 guards outside on the metal frame of the storm window. It is nighttime vs. the daytime when they posted no guards. Those guards will probably be frozen to the metal frame like the ones before and then I won't be able to thaw them! There are 8-10 bees nursing the bottle.

October 18, 1983

Last night after a New England boiled ham dinner (one of my favorites and I make navy bean soup out of the bone and bits of ham), I was watching the hive. A bee appeared carrying in her mouth a white/grey thing. It was a ghastly sight. On close examination it may be a drone larva. It had definitely been uncapped. It was in an advanced enough stage of development that I could tell it was perhaps a drone.

My supposition is:

It's cold now---32 degrees. Days are shorter. The nectar flow is drying up. Pollen is scarce. They are aborting the drones because they won't carry them over the winter. Another instance of "Drone Slaughter."

Several days ago I noticed a number of dead drones on the stones outside the entrance to the hive. I imagine they went out to play and when they came back, thousands of their sisters were guarding the entrance and wouldn't let them in. So they froze to death. Also, I have noticed recently that the drones who "beg" for food have been denied by the workers. It is ultimate pragmatic time in the hive. They are always pragmatic, but going into winter, they become more so. The drones are unproductive and must go. "No Work, No Eat" is the winter bee's slogan as it was during the dire straits of Jamestown.

My bees are sweet and produce sweets, but they have a sting---are still part of the Fallen World. Abortion, murder, exposure are integral parts of their world as they are in ours.

The worker bee paraded that dead drone larva (?) just like they do the other dead. It is a ritual, this Parading of the Dead.

I continue to find hive debris on the inside sill here in the kitchen where they push them out---somehow---only the Lord knows how. There are lots of dead teeny ants, wings, legs, pieces of wax.

November 19, 1983

I am really furious!! When I put in the sugar water, c. 8 drops plopped out onto the hive floor! I have found so many defects in the design of this indoor hive!! I've had to watch this bottle like a hawk. I turn it upside down, create the vacuum and put it in the hive. Then, always, out plops 5-8 drops of liquid. When I think of the thousands of my bees who died because of this design flaw and the moisture design flaw, it makes me LIVID!

When I saw the leak, I took the bottle out, plugged up the hole with the clay and watched and waited. Within 30 seconds, a worker hung from the bottom of the brood comb, put out her red nectar tongue and began to absorb the sugar water from an upside down position. Presently another appeared on the hive floor---oops, her wings just got caught in the sugar water.

Watch out for that deadly stuff that oozed out into the other hives and killed your cousins. Why do "they" keep telling me to give them sugar water?! (Why do I keep doing it?! Because the experts tell me to do it. You know better than that, Sandy. Or do you?) She begins to sip, also. She leaves and regurgitates it somewhere. She and two others are back sipping, their red tongues out, their little shiny black bottoms pumping up and down to beat the band. I always take this pumping of the bottoms to indicate enjoyment.

I have only seen this rear throbbing, pumping when they take sugar water, get nectar from a flower or sip honey from hive comb. So I think it is associated with eating. It probably is physiological. The pumping of the abdomen has something to do with the ingestion process? Or is it pure pleasure?

Now, 10 minutes, later, there are 12 bees sipping. Five are hanging upside down from the brood comb (the smart ones, they aren't going to get icky) and 7 are around the puddle. They'll clean that mess up in no time! The hive has come alive---sipping, running back and forth. I love to see the bee activity. This particular hive has been so static on this side due to the bee gate reversal talked of earlier. The only real excitement has been the trauma of the sugar water bottle! (Don't always pay attention to the experts, Sandy, especially if empirical evidence and logic contradict them!)

November 24, 1983

Thanksgiving Day. I have a turkey stuffed with dressing browning in the oven. Made candied sweet potatoes. Will have mashed potatoes and gravy, cranberry sauce, black and green olives, peas, pumpkin and cherry pies and a chocolate cake. Meal fit for a king. How blessed we are!!

Thank You, Lord, for all our Christian brothers and sisters who are right now feeding such a dinner to tens of thousands of people in homeless shelters and churches!!

Now I will give my babies some fresh sugar water. It's warm, 58 degrees, foggy, rainy and they're running around.

Several bees are dying today. One is being paraded around half-alive.

Yesterday I saw a very small fly-like insect on the glass and inside the boards of the hive. It was much smaller than the little flies who were allowed to live in the hive for so many months.

Now she's dropped the half-dead bee on the floor of the hive. The dying one is going about the business of dying.

Just saw a little ant crawl over the floor of hive, up the glass, down the glass and away---she's gone out of sight. How many different kinds of life are living in there now! Not good!

December 12, 1983

Note inserted later: Because I entered nothing in this journal other than this date, I was obviously interrupted in my observations. Hopefully, to listen to Christmas carols with my dear Blake and Jesse. "Ding-ding-a-ling; Ding-ding-a-ling." And my Dee (Kathy) would have been coming home soon for Christmas.

January 3, 1984

Last night I had Steve turn the hive around so I could see the outside comb. It was empty. They're in the inner comb with their stores. They have laid no honeycomb on the outside, but they have laid brood comb and brood had been reared there right smack in a circle in the middle of the brood comb. All the rest was unlaid comb. Just a circle in the middle of brood comb and a larger circle totally empty. Centrists again.

On this side, they've begun comb in a similar pattern. There is comb starting in the middle and making a giant "C" up to the top of the brood comb. All around is barely laid comb, again in a broad "C" pattern.

There is a lot of dwindling. They are dead and curled up on the bottom of the hive. Interestingly enough, most of the dead are on the kitchen side of the hive. When we looked at the other side, there were only a few dead there. Go to the warm to die? I wonder if they go to the warmest place to die in a wild hive?

Maybe the long brown streaks on the glass and on the combs are feces. I get a lot of dots of those on my car in the summer. They're not able to go out so they void themselves in the place LEAST likely to affect their actual living space---the glass panel and unoccupied cells.

I have not been giving them sugar water. The Battle Of The Bottle is too fierce. Since I can't see their stores, I can only hope they have enough.

Steve got me a good Nikon camera for Christmas. I'd like to get into microphotography and really see my bees.

I'll hear about <u>The Heart of the Honeybee</u> sometime in January.

January 11, 1984

Note inserted later: Just read about the most amazing beehive. It's on the roof of the Opera House in Paris!

In this year, 1984, Jean Paucton, a backstage worker in charge of set furniture for the Opera House, bought a hive of bees and was keeping them in his apartment until he could take them to his place in the country. He brought them to the Opera House and installed the hive on the roof of the Opera House. A week later he checked to see if the bees were okay. The hive was full of honey! The bees had foraged for nectar in window boxes and in the Presidential gardens up and down the Champs Elysees. Over 200 lbs. of honey comes from that one hive. Apparently the honey is very "strong," but it is sold in the souvenir shop of the Opera House and at Fauchon's, a trendy food store in Paris. It's called "Opera Bees Honey." Apt name, because honey soothes the throats of opera singers before they fill our ears with those "mellifluous" arias! (Latin "melli" meaning "honey, sweet.")

The interior view of the hive in the dead of winter is a grisly one. The first impression---death. Winter dwindling leaves the dead curled and drying all over the hive. They mostly go to the bottom of the hive to die. Reminds me of the old wive's tale about dying people working themselves to the bottom of the bed. I can attest to this. When Grandpa who was 92 was dying, I found him scrunched at the bottom of his bed for several days before his death. I would think of this old wive's tale as I would pull him back up and prop his head on the pillow. Also, I go several days a week for years now to the old people's home, Altna Craig. I've witnessed several do this for days before they subsequently died.

There are several dead at the bottom of the honeycomb and a few dead on the brood comb. They cling to the cell for several dead days and then fall to the bottom.

There are long streaks of auburn feces on the glass and feces on the empty honeycombs and on the brood comb, too.

Minute drops of precious propolis, the bee's glue, are randomly smeared on the wood. In contrast, propolis is purposefully and meticulously packed along the sides of the hive---geometric lines of it against disease, cold, moisture.

Small bits of that granular white stuff cover the relatively clean floor and are packed against the glass between the honey and brood comb---out of the way.

There appears to be no life, but in reality the queen has begun laying eggs for the early spring bees and the sure nectar flow. Shakespeare's Appearance Vs. Reality theme.

No hum. No noise. Soundlessly their winter life goes on. The same life that waits surely and soundlessly beneath the snow, as beats and abides under the ancient copper beech tree out my window that now is draped with 8" of fresh snow. Under this great copper beech are thousands of tiny, persistent scilla siberica bulbs even now multiplying and dividing to the infallible arithmetic of God's Plan. He, as the Loving Mathematician, combing numbers and the Nouma, the Holy Pneuma, in all, through all---is all.

So the deadly mess will go away, as our deadly messes eventually do. And in its place, because of the Pneuma, will be life and order and production and profligacy and possibility.

A loud Molly Bloom "Yes" to profligacy and possibility!

January 19, 1984

I hope they are alive. I've had the window open for 30 minutes and no bees---only the dead on the hive floor and a few on the queen excluder. In hope, I made some sugar water and have attached it so it doesn't drip into the hive. It's been 15 minutes since I put it in---no takers.

Last night late I had lots of fun making up Bee Limericks.

> "There once was a drone from Bayonne
> Who had sex with a queen from Lyon.
> Instead of choice nectars
> And pollen from hectares
> Their brood gathered gateau and pones."

I wrote about 8 or 9 of them. Cute and fun.

Hope I get the Workman book!

January 25, 1984

In spite of my optimism several weeks ago, I believe all are dead in there. Again. Perhaps they starved to death. Or this damnable feeder kills them by dripping or kills them because they don't have it. Or?

Today I heard Sally at Workman's didn't go for the drone chapter. I didn't think she would. She TOLD me in the first letter she wanted it to be "quirky and informative." That is what I did, but when we talked to her, I saw that's NOT AT ALL what she wanted. She really wanted a bee book a la <u>In And Out Of The Garden,</u> Sara Midda's exquisite book. That's what I wanted to do. Steve got her to let me have another whack at it. I'll do that but, Jesus, You have to help me. Write it through me. I've got the material in me. I've done my work. Please do Yours.

April 27, 1984

Took me three months of work, but I sent in 15 pages of <u>The Heart of the Honeybee</u>. Thank you, Jesus. You did give me so much of the illuminated design. I had such joy doing it. It's in Your Hands completely, thankfully.

Maybe You just want me to write and never be published. That's okay, too. The great joy for me is in the writing and research anyway.

April 29, 1984

As I had suspected, all the bees died in the nuc hive. Today I picked up and hived my 5th package of bees in two years. I took out, cleaned out, scraped (with my trusty hive tool) and fixed all the interior of the hive in preparation. That takes loads of time. Usually takes me over 24 hrs. of really hard work. Except for the first hiving, all of the others have been uneventful, smooth and very exciting.

We, Steve and I, did it again. He's overcome his fear and helps enormously. I've got so I'm not very apprehensive. I'm bolder but still cautious. There are 10,000 stingers in there! But I know, have faith that they are not going to sting me. They just want to live and to get in a home.

But the hiving now, as with my Christian faith, is built on pragmatic experience. The pragmatic experience is hiving them again and again and living with them day by day. My Christian pragmatic experience is almost 30 years of living with Him and allowing Him to home in my heart.

It's now 1:00 at night. I let them run in for 9 hrs. After the Bible Study, I closed them up. Steve and Blake brought them in and installed them in the window. (Those two who attacked the "killer bees" years ago on Poplar Rd. and opposed my getting bees have become my mainstays.)

The glass was steamed up and during the moving of the hive, the honeycomb was moved too close to the glass and crushed several bees. I reopened the hive, went in and positioned the honeycomb correctly. The bees didn't even whirr.

The queen cage is on the left side. They're all clustered around it. They're eating away at the candy plug that is substantially bigger than the other plugs have been.

I just talked to Kathy at Pitt for an hour. How wonderful she is and is doing so well at school. She's running that school as she will everything in her life.

Introduced the sugar water feeder. It seems, miraculously, to be working. They're nursing it.

I praise You, Lord, for the honeybee. They're more than and the same as the Blakean "World In A Grain Of Sand." William Blake was such an influence in my early intellectual life. When I did my thesis on <u>The Four Zoas</u>, I wanted to make his prophetic works accessible to all. The thesis, of course, ended up being as Blakean as could be! Failed there. Since William Blake and his wife Catherine didn't have any children, I named my dear Blake after him because the poet had given me so many "mental" children!

Truly Your design reflects, gives light into, indicates You, the Designer. There is intellectual power in the teleological argument for Your Existence.

The man who put honeybees "on the map" scientifically was L.L. Langstroth, a minister who was a beekeeper and a scientific observer of them. His book <u>Langstroth On The Hive and the Honeybee, A Beekeeper's Manual</u> (1853), radiates the same awe of the honeybee and the same awareness that I have. That is: this microcosm is a teeny, teeny reflection of the huge cosmos that God has created. Every thing, animate and inanimate, that one can delve into is filled with complexity and fascination. There is the possibility of infinite research into any thing! Just try to understand yourself whom you live with and converse with every nanosecond! I love the cosmos within as well as the infinite multiplicities without! Thank you, Lord.

Here's Langstroth's reason for writing the first comprehensive book on beekeeping:

"I have determined, in writing this book, to give facts, however wonderful, just as they are; confident that in due time they will be universally received; and hoping that the many wonders of the economy of the honey-bee will not only excite a wider interest in its culture, <u>but lead those who observe them to adore the wisdom of Him who gave them such admirable instinct.</u>"

I love how Rev. Langstroth wanted his research to lead people to the Lord!! Yes, everything we do should take the focus off of us and put it on our Creator!

And I love another minister, Reverend Spooner. Found another great Spoonerism. (Our family laughs at Spoonerisms all the time!) At a wedding he said, "Son, it is now kisstomary to cuss the bride." Rev. William Spooner (late 1800's I think) was an Anglican priest at Oxford. He always inverted words. I sometimes say, "weans and beiners" when I mean "beans and wieners." He did that ALL the time. So frequently and publicly did that type of verbal gaffe spill over into his public life that it is called a "Spoonerism." In lifting his glass to toast Queen Victoria, he called out, "A toast to our queer old dean." I find these hysterical. In ushering a woman to her seat in church, he said, "Mardon me, Padam. May I sew you to your sheet." Legendary mix-ups in chapel: "Our Lord is a shoving leopard" for "Our Lord is a loving shepherd." "Which of us has not felt in his heart a half-warmed fish?" He was trying to say, "half-formed wish." When he visited a friend's home in the country, he said, "You have a nosey little crook here." He reprimanded a student for "fighting a liar in the quadrangle." "Lighting a fire in the ..." He was a brilliant scholar, but he was definitely tongue-challenged and an absent-minded professor. Once he invited a man to afternoon tea "to welcome our new archaeology professor." "But, sir," said the man. "I AM the new archaeology professor." "Then never mind," said Spooner. "Come all the same." God bless Spill Booner!

April 30, 1984

These new bees are fantastic! Just like my first bees. The hive is clean. They have filled the brood comb with nectar. They have freed the queen in record time. They are happy. I am happy just sitting in...

Just got a call from my dear Kathy. I'd interrupt any study for one of my loved ones!! I love, love to study, but I forego that all the time in favor of my dearest ones. Frankly, I find them much more interesting and more satisfying than I do bees, or illuminated mss., or horticulture, or Jung or any of the many things I get into.

May 1, 1984

May Day! But there is no emergency in this hive. Only the urgency to get the brood comb ready for new life. "May Day!" used as an emergency call actually comes from the French "M'aider" meaning "Help or aid me." It's not correct French, but it's been used for over 100 years.

Interesting---the 11th Space Shuttle that went up in April contained an experiment on whether the honeybee could construct honeycomb in zero gravity. The little ones went to work and laid comb just like they do here on earth. I think I read somewhere that insects are not bound by gravity. That's why they can walk on ceilings. So maybe they were trying to figure out what would happen to wax and wax structures built in zero gravity? Or maybe they wanted to know if they COULD excrete wax in zero gravity? Or?

Re: comb. The few honeycombs that the last "nuc" bees drew are the most beautiful I've ever seen or probably ever will see. Fat, puffy, white---each cell is very, very deep. When a bee goes in, her whole body disappears and there is still room to spare. They didn't draw much comb, but what they drew was nonpareil. They are, of course, hexagonal on the inside, rounded at the entrance, but these are deeper, sharper, more perfect, gorgeous!!

I read that the hexagon shape is the most efficient of all possible shapes for storage capacity and reuse. The square and triangle are close seconds. In the wild the honeybees instinctively make all their hive cells hexagonal. And in those hexagons they store honey, pollen and developing brood. They utilize the hexagon for most of the products from the hive including the main product---themselves.

An observation that I've made repeatedly: One bee will go throughout the comb and will stop at bees here and there. She will grab the head or the back leg of the other bee with her front legs and will wiggle her body up and down. Then she leaves the other bee. She has completed whatever it was that she was doing. She moves to another select bee and does the grabbing, shaking and leave-taking. Is this an in-hive Dance? Shaking Dance. One has the feeling that she is a messenger sent to certain ones only. She is communicating a task, perhaps, or...

May 16, 1984

I haven't written in this log for a while because I've used this book to hold open the top cover of the hive. There is a top cover on my observation hive, then a deep space and then the inner cover. If the inner cover is removed, the bees can escape into the kitchen. Yes, there has been condensation in the hive and I'm trying this new thing with the book to see if it helps. The condensation on the glass appeared to clear up with this top propped open.

These bees have done an interesting thing. The honeycomb that was most beautifully wrought by the doomed "nuc" bees was too DEEP for these bees' taste. So they have stored the nectar on the brood comb. Meanwhile a destruction/construction crew of workers pares down the deep hexagons made by my nuc bees to their own liking. The crew has been hard at work on the right side of the honeycomb and has taken off about a half inch of its depth. I mentioned that it was so deep that the bees were totally encased once they went inside. It's not at all unusual these days to see a bee with a big, half-inch chunk of wax in its mouth meandering here and there.

A Cosmos in my Kitchen

They are storing nectar in the brood comb starting at the top of the comb. The whole top is covered with nectar down about 4 inches PLUS the right hand side continues on down about another 2 inches.

They are, of course, packing in the pollen. It is scattered here and there. Looks like earwax (aka "cerumen" from the Latin word "cera" meaning "wax").

I loved taking Etymology in college! I ended up with a double major in English and Classics. Professor P. was one of the three professors in the Classics Department, so I took a lot of classes from that very dapper, short, middle-aged man. I always got "A's" in his classes. In the middle of my junior year, he told me he and some other professors were recommending me for Phi Beta Kappa at the end of my junior year. I was thrilled. But I got a B on one of his tests. I didn't get Phi Beta Kappa until my senior year. Professor P. told me it was he that had recommended they wait until my senior year. I always respected him for that. There is a mark and I had missed that mark. In my senior year in college, he bought a Mercedes sports car. One day I was walking along near the Carnegie Library and he honked. Nice car, I said. Get in, he said. We drove around Schenley Park and then he dropped me off. After graduation, I got a Teaching Fellowship and taught at Pitt for three years as I was getting my Master's Degree. He and I were good friends and would often have coffee together. One day he appeared outside my classroom. "I have to talk to you immediately," he said. Over coffee, he told me that he and one of my sorority sisters were having an affair! On the outside, I remained calm, but inside I was nonplussed. He was married to a very nice looking, petite older woman. The long and the short of it is: he and my sorority sister had a year-long affair. Her parents found out and she withdrew from the University. Professor P. and his wife did not divorce. He and I would have coffee occasionally, but we both felt "awkward" (from the Old Norse "afugur" meaning "turned backward" and the O.E. "awk" meaning "back-handed"). What he/they had done was so "sinister" (from the French "senestere" meaning "left" and the Latin "sinister" meaning "left" and the English "sinister" meaning "bad, evil, base, wicked"). And the definitely not absent-minded professor had acted like a real "goof" (from the English "goff" meaning "a foolish clown").

They are preparing the remaining brood comb for eventual brood. I realize that some of the nectar is in the brood comb for the brood, but not this much.

I haven't seen the queen since I hived her, but I know she's there by the work activity. This is a happy, busy, well-run hive.

Talk about the queen, I just saw her long abdomen. She's lean and hurrying over to the other side of the brood comb. The white dot that marks her back is almost gone. Probably from so much grooming!

There are about 14 drones lolling on the upper left hand corner of the brood comb. The women are bustling around between and over them.

May 17, 1984

I just took this diary off the top cover, closed the cover gently. My---what a flurry of excitement! They are SO sensitive to any jarring or touching or disturbing of their city.

All bees respond to the subtlest of stimuli. Knock the hive---flurry and buzz. Open the hive---flurry and buzz. Close the hive---flurry and buzz. Even when I put my nose NEAR, not on, the side vent to smell inside, several bees go zzzzzz.

I love the hive smells. My favorite is, of course, the spicy, hot smell of cinnamon. I've concluded it's a certain type of propolis because the hive that smelled most like cinnamon was the one whose propolis tasted most like cinnamon. Some of the other propolis I've tasted is not as flavorful and aromatic. This smell filled my kitchen for about 4 months with my first bees who really loaded the hive with this propolis.

I, also, love the heavy, sweet smell of the honey ripening in the comb. I've mentioned the slightly fermented smell it has from time to time. I like that, too.

I read up on that fermented smell because I didn't think honey could "ferment." Of course, I knew of mead and the drunken mead halls from <u>Beowulf.</u> (The Old English name "Beowulf" probably means "Bee Hunter.") In that epic poem I first learned that honey can be fermented by adding yeast to it and made into a powerful alcoholic beverage, mead. But I didn't know honey could "ferment" in the hive! Mead is man's oldest alcoholic drink. Of course, crude beers made from grains are very, very old, too. In order to get the fermented honey drink, humans have to MAKE it. The fermented smell in the hive was definitely not bees making hooch. In my buddy Langstroth's very erudite and scientifically phrased book it says: "Fermentation (of honey) is caused by the action of sugar-tolerant yeasts upon levulose and dextrose, resulting in the formation of alcohol and carbon dioxide." That's how it happens, but why did this happen within my hive? He goes on to explain that all honeys contain yeasts. The yeasts are from the nectar in the flowers! There it is. Nectar is gathered by honeybees. When the bees extract the nectar from the flowers, they give it some of their enzymes. The nectar is stored in the hexagons. The nectar loses its water content and gradually becomes honey. But it is occasionally contaminated by some of the yeasts that are found IN THE NECTAR that they bring into the hive. Long-winded but, I think, accurate description of why there is a fermented smell.

Even though it may ferment a little, liquid honey does not ever spoil. It may become granular but the high sugar concentration in honey kills all bacteria. Of course, the ancients knew that fact because for thousands of years they used honey to embalm and preserve their dead.

I remember when my book group did John Gardner's excellent little book <u>Grendel</u>. Steve and I read it out loud to each other and were delighted by its language and intelligence. For years Steve would look at me and go, "Waaa, Waaa" like Grendel did in the book. Of course, the book was a brilliant rift on the Old English epic poem <u>Beowulf</u> in which Grendel is the monster slain by the hero Beowulf. I have my <u>Beowulf</u>. I'm going to write down the first mention of the "mead hall" (fermented honey drink) in the epic and then the modern English translation.

<u>O. E.</u>

"Him on mod bearn,
paet healreced hatan wolde,
<u>medoaern michel</u> men gewyrcean
ponne yldo bearn aefre gefrunon,
ond paer on innan eall gedaelan
geongum ond ealdum, swylc him God sealde
buton folcscare on feorum gumena."

Modern English

> "It came to his mind
> to bid his henchmen a hall uprear,
> <u>a master-mead-house</u>, mightier far
> than ever was seen by the sons of earth,
> and within it, then, to old and young
> he would all allot that the Lord had sent him,
> save only the land the lives of his men."

Old English (O.E.) is truly a foreign language! The underlined is the mention of the "mead," the fermented drink from my little bees. The events in <u>Beowulf</u> take place in the 400-500's A.D. The poem was written down centuries later by monks in England. Most scholars think the fabric of the story is a valid representation of the mores of the time. In those very early days of England and the English language tribal kings built mead halls for themselves and their soldiers to congregate. We can see from the word "mead" that the strategy sessions soon turned into drunken brawls. It is said in the epic poem that the mead-benches upon which the men sat were bolted to the wall. Of course. There is, also, a mention of "beer" in the poem. So man has had beer and mead wine forever.

Just had to go and get Chaucer's <u>Canterbury Tales</u> to show how Modern English evolved from <u>Beowulf's</u> <u>O</u>ld <u>E</u>nglish into Chaucer's <u>M</u>iddle <u>E</u>nglish. I love and memorized this Prologue from the <u>Canterbury Tales.</u>

M.E.

> "Whan that aprill with his shoures soote
> The droghte of march hath perced to the roote,
> And bathed every veyne in swich licour
> Of which vertu engendred is the flour."

Now by the time of Chaucer (c. 1343-1400 A.D.) 800 years after <u>Beowulf</u> we can begin to understand some of Chaucer's words. The modern English translation is:

> "When that April with his showers soote,
> The drought of March has pierced to the root
> And bathed every vein in such liquor
> Of which virtue engendered is the flower."

"Soote" means "sweet." I could go on forever.

Later: The queen is laying now. She walks around. She pokes her head in a cell. She enters it with her head and thorax in the cell. If it looks good, she withdraws, advances several cells until her abdomen is over the previously examined cell. She inserts the abdomen. She swirls around making a semi-circle. Then she braces herself with her legs on the adjoining cells and lays the egg. When she is finished, she withdraws her abdomen and moves on continuing her search for what she considers an appropriate hexagon.

It must be noted that she does not ALWAYS make the semi-circular turn. But I have observed this at least 50% of the time.

She laid on this side for 1 hour. She has exited this side on the bottom right

May 20, 1984

I have just been observing the different stages of developing brood before capping.

The earliest stage I've been able to see is one white, teeny stick-like being on the polished surface.

In the second larval stage, the stick starts to fill out and curl into a "C."

In the last stage, the larva keeps filling out and develops into an "O" which fills the cell. This is quite a beautiful stage of larval development because the fat larva is white-white and glistens.

May 22, 1984

Ciel R. and I were observing the hive and a bee appeared with a larva in her mouth. She was parading it. Did a robber get in and take it out of its cell?

May 27, 1984

We had a wonderful Bible Study tonight in our Carriage House. We all have met here every Sunday evening for five years now. There's no schism in me between the world of my Father and His children and my everyday life and my bees. It's ALL holy! Of course, I'm so fallen that it's not ALL holy!

I saw something I never read about or thought could happen. Two (2) larvae are developing in the same cell! Twins. There's nectar on the bottom of that cell. The bees kept covering it with their bodies, but I stood fast and have observed them for about 10 minutes. I'm ABSOLUTELY positive that I see 2 larvae in that cell. I left to get a piece of tape to mark the cell for continual observation and now I can't find it. I know the exact angle of about 15 cells so I've marked them and will look tomorrow.

Note inserted later: Twins are caused by one sperm fertilizing one egg. Within 15 days that egg splits in two. The resulting identical twins share the exact same DNA (have 100% the same genes). There is an on-going study of identical twins at the University of Minnesota. In 1979 Thomas Bouchard came across a set of identical twins, "the Jim twins," who had been separated at birth. Their similarities were so striking that Bouchard started The Minnesota Twin Project to study the result of Nature vs. Nurture on twins separated at birth. The "Jim twins" found each other when they were 39. Both had been named "James" by their adoptive parents. Both were 6 feet tall and weighed 180 lbs. They both smoked Salem cigarettes, drank Miller Lite beer, had sons named "James Alan," had named their childhood dogs "Troy" and held part-time jobs as a sheriff. Amazingly, each had married twice and each first wife was named "Linda" and their second wives were both named "Betty." Each twin had driven a light-blue Chevrolet down to Pas Grille beach in Florida for family vacations. (I wonder if they ever passed one another on the street?) Of course, the Minnesota Twins Project raises questions about the effect of the environment on people. The age-old fight between Nature and Nurture. My mother and I always haggle over Heredity vs. Environment. I tend to lean more to environment. She is firmly in the camp of heredity. "They never ran a jackass in the Kentucky Derby," she is fond of saying at the end of our discussion. "You can't make a silk purse out of a sow's ear" is another of her rejoinders. Pavlov with his salivating dog and B.F. Skinner in behavioral psychology have permeated academia with a bias toward Nurture. But the twin studies at the University of Minnesota do give pause---a big pause.

There's a town in Ohio near the Akron/Cleveland area called Twinsburg. Identical twins, Moses and Aaron Wilcox from Killingworth, Ct., purchased 4,000 acres of land in the area in 1819. They offered six acres of land for a public square and $20.00 to start a school if the settlers would name the town Twinsburg. At least they were not name-vain---Wilcox, Ohio; Wilcoxville, Ohio; Wilcoxburg, Ohio; Wilcoxton, Ohio; Wilcoxland, Ohio; Wilcoxborough, Ohio. Long before the Minnesota Twins Study uncovered the uncanny likenesses of identical twins, the residents of Twinsburg, Ohio remembered that the Wilcox brothers held all their property in common; were business partners; married sisters; had the same number of children; and died within hours of each other on the same day of the same fatal illness. They are buried together in the same grave in Twinsburg. Their identical legacy goes on. Each year since 1976 Twinsburg has hosted a reunion of twins from all over the world.

We've been summerizing the pool area. The honeybees are swarming all over the maple tree and to a lesser degree on the tangles of honeysuckle. For three days now they have been so thick on that old, tall maple tree (getting propolis?) that there is an audible hum all around the area. They flocked to the maple the day after it shed its thousands of chartreuse clusters. The driveway was a carpet of threads. I guess that shedding is the signal for heavy nectar/propolis/pollen flow?

Or---I know that before most plants bloom, they have traces of propolis on them. Maybe they are getting propolis? Every day I see the ants and other teeny insects getting propolis from the buds of my gorgeous and reliable peonies. Love peonies. If you have one, you'll have her for life! There are peony bushes in Japan that are 300 years old! Now that's a reliable plant!

Every day now one or more of my bees buzzes me around my head to say "hi." I often put up my hand and she touches me. Then she's off. Just a friendly howdy-do. Love them! Thank you, Lord, for the honeybee.

May 28, 1984

Not one. Not two. Not three. But FOUR CELLS IN THE SAME SMALL AREA HAVE TWIN LARVAE IN THEM!! Uh, oh!

As I am looking in amazement, a bee emerges on this side parading in her mouth a fully developed WHITE bee that they have, no doubt, aborted. I'm sure it is a hive bee that aborted this larva because it's been cold and raining all day and there are no bees or other insects out to enter the hive and pull the larva from its cell. Then another bee appears carrying another aborted WHITE (sans colorful markings) larva. So:

1. Possibility that this queen's ovipositor malfunctions occasionally and drops two eggs instead of one into the cell.
2. The workers "know" that the hexagonal cell is static. It's not like an animal or human womb that is elastic enough to hold many beings. The cell can really only contain one fully developed bee. Hence, I conclude they are aborting the capped "twin" cells at an advanced stage of development.

The WHITE, ABORTED LARVAE have the normal three sections to the body. Everything on them is white except the eyes that are brown. The white bees look complete to me minus the distinctive coloration that the emerging bee has. They're completely developed but are still in the white larval stage of coloration.

They've been parading (another Parade) these white bees before me for 15 minutes now. I notice they have concentrated the parade on the section where I espied the first four sets of twins---coincidence?

Are these twins honeybee larvae? Or?

A Cosmos in my Kitchen

June 22, 1984

I've been so busy the last three weeks that I missed the swarm! It's obvious that a host of them has left.

This side has a hundred or so bees vs. the usual several thousand. There are sporadic capped bees (c. 40). The honeycomb is replete with uncapped nectar. The smell of the ripening honey permeates the kitchen.

Now that most of the first brood is gone, I see that the new brood is laid in the same circular pattern that I've observed before. Centrists again. You did a nice job, clipped queen. I'm sure they killed you and built another queen to lead them out on that swarm.

Now that I've observed long and closely, there seems to be something wrong with the developing bees. The caps have been torn off of some of the brood. One is inside the cell, black and dead. Another still inside the cell is dead, cap removed to expose her deadness. I notice holes in the caps of some of the other brood. No activity. SOMETHING IS WRONG! Not again!!

Also, the remaining bees are dispirited. Might be the time of night---it's midnight---or something else.

The brood that remains is NOT normal. Maybe you didn't do such a great job, clipped queen?

July 15, 1984

I think I've got a queenless hive!

Few bees. They go in and out, but I don't see any pollen being flown in. Dead bees are left for days on the bottom of the hive before they are removed. The hive has become host to other insects: black ants, tiny black flies and a particularly disgusting, white, segmented worm with one black-red eye! Could that worm be the cause of the irregular cells I've seen? No, please. What is it!?

The honey is over-ripe. Should have been capped. The fermented smell is all throughout the kitchen. Glad you can't get drunk from just the smell of alcohol. In ancient Greek the word for "drunk" means "honey-intoxicated." I am definitely a honeybee-intoxicated person.

The defective brood, uncapped and partially capped, has not been removed and remain dead and black in their womb-graves.

It's---all told---a holding pattern hive. It is still life sustaining but is not meaningful, productive, purposeful.

Like so many lives outside of the Lord. Just THERE but not really ALIVE.

My current estimate is: in June they swarmed with the clipped queen. She fell to earth, of course. They smelled her absence, returned to the hive, hung around a while and left en masse to find a new queen or left with a new queen or left to build a new queen elsewhere. They wanted out. What do they know that I don't?

Amidst all these failures I did find a cute poem on Swarming:

> "A swarm of bees in May
> Be worth a load of hay.
> A swarm of bees in June
> Be worth a silver spoon.

Swarming in July.
Let the buggers fly."

Meaning that if they swarm in May, that's okay.
If they swarm in June, that's a boon.
If they swarm in July, who cares?

The skeleton crew that remains in the hive has no fertilized gyne (queen) cells, can't build another queen and are existing there because they are there and it is their home. I estimate there are several hundred in there. The prognosis is not good.

Well, well. I've had so many failures. But I've learned so much from the failures that I wouldn't have learned from successes. Everything here is analogous to my life.

Tomorrow I should hear whether Workman will do <u>The Heart of the Honeybee.</u> If so, Lord, and I would like it to be so, please have it glorify You through these refulgent creatures.

Amen.

July 25, 1984

I just had my first "verbal communication" with my bees!!

I was observing them for a long time. (There MAY be a queen there.) I closed the hive and this incredibly loud humming started. It was not the hum of many bees, but the hum of one bee who was really turning up the volume.

I put my mouth against the vent and tried to make the same sound I heard emitting from the hive. The long and the short of it is: For 15 minutes this one bee and I "buzzed" each other. She would do a noise. I'd imitate it as best I could. There would be a pause. She'd do another noise. I'd "answer." Pause. She'd "ZZZZZZZZ." I'd "ZZZZZZZ."

Blake, Ainsi (another au pair from Sweden), Paul M. and Chris R. were all here in the kitchen and all heard this communication! They, of course, think I'm "eccentric" buzzing and humming on my knees into a tiny air vent in a beehive. But they heard her buzz and me answer. I know we were communicating in some way!!

My bees have opened me up to other insects. I got a little bottle with a red nozzle to try to attract hummingbirds. Haven't seen any yet. But I did think I saw several hummingbirds on the huge stand of red and pink bee balm. Instead I found out they were hummingbird moths!! They are gorgeous, dainty, and to my untutored eye, they look exactly like the real thing---hummingbirds. The hummingbird moths love that bee balm and fly from big flower to big flower and never alight. They just hover, hover with wings going a mile a minute getting some of that good old Oswego Tea!

I have put out my palm under them and they have permitted it. As they sip, I hold my whole hand cupped around their tiny bodies and feel the faint but incessant whirr of the delicate wind of their wings. I talk to them and thank them for the privilege of being so near.

Also, the big, fat bumblebee lets me pet his/her furry thorax. I am somewhat fearful as I reach out with my index finger to his thorax, but I am determined to pet him and he probably thinks it's a novel thing for him, too.

The stand of bee balm is awash all day with bees, teeny insects, the hummingbird moths and lovely, skittish butterflies. They haven't let me pet them, though I have tried. This magnificent specimen of balm is about 5 ft. in circumference and very visible, I'm sure, from the air. But my honeybees don't come to it. They prefer the smaller flowerlets of the nearby pink liatris.

Just read an article on insects. Seems most of them are congregated in the Amazon. The man wrote that 75% of the combined weight of all insects in the world is in the Amazonian forest. Insects do love and thrive in the heat! When I spent that summer in the Yucatan peninsula studying the Maya, I had to sleep under mosquito netting there at the mission in Merida. I'd get into my nightgown. The mosquitoes and other insects would be all over the room, the cement floor and the, aptly named, mosquito netting that hung over my little, low twin bed. I'd lift up the bottom of the netting very carefully until it reached the bed. Then I'd try to jump in and drop the netting to the floor again in one fell swoop. Since I was in the Yucatan for so long, I became adept at that maneuver. Every morning when I would open my underwear drawer, hundreds of cockroaches would scurry away. I'd turn the panties inside out and wear them anyway. Lots of insects in the world. But there were bigger things to worry about in the Yucatan. All the missionary children had tapeworms. Most of the missionaries had or had had all kinds of fevers, parasites and diseases. They would go to the Mayo Clinic when they had their Sabbaticals and the Clinic loved to see them arrive. The doctors learned so much about exotic parasites from studying them. I shared the little house on the grounds of the mission compound with a woman who was 82. She had come to the Yucatan as a missionary when she was 21. She told me that when I went out to the ruins at Chichen Itza or Uxmal or wherever, I should put soap around my ankles, arms, neck or any place that had an opening where chiggers could get in. Chiggers are teeny mite-like insects that hitchhike on humans. I dry soaped up every day just like she showed me. Never in the summer I was there did I get a chigger or any other kind of insect in my skin. Miss B. had never married. She had lived with and loved and cared for in the Lord the Yucatecans for over 60 years. She had no family in the States and no home to go back to. She had elected to stay with her people there and to be buried in one of their graveyards.

As my mother always says, "There's many ways to live a life."

September 19, 1984---Block Island, Rhode Island

Steve and I and Deanna and Roger have come to Block Island, R.I. for several days. We have a nice beach house. At this time of year Block Island is very deserted. We have the little town, the restaurants and the beach to ourselves. Went out for ice cream after dinner yesterday and in addition to the four of us and the scooper guy, the entire senior class of Block Island was in there. That made SEVEN (7) people in the shop. Yes, the senior class is composed of 2 people. I asked them: "Well, whom do you marry?" They said, "Summer people."

With the Ciskies we are with good friends, have good Bible Studies, play Bible Trivia Pursuit game, eat great food---cooked in and out-—and have a chilly sea breeze against our faces and backs as we are pushed along the rocky shores. Today we went to the head of the island where a marker has the list of original founders of the island and there was a Sands among the small list. Meme was a Sands. Bev C. in our Bible Study said he is a Sands, also, and produced a book on them showing their wanderings and descendants. So Bev and I are very, very distantly related and I am very, very, very distantly related to that old salt (for he must have been to inhabit that aerie) who was one of the founders of Block Island.

At 2:15 this afternoon the four of us were walking on the beach. I heard a lot of furious buzzing and went to find out what was going on. Another mating of bumblebees.

The conditions were:
1. Gorgeous, sunny day.
2. 75 degrees.
3. They were mating IN the wet sand. They were both semi-wet and matted and sprinkled with sand.
4. They were c. 2 feet away from the encroaching surf.
5. 2:15 in the afternoon.
6. They were in full view on the empty beach, not hidden among rocks.
7. It was windy. Not excessive wind, but enough to roll them around occasionally.
8. They were about 20 ft. away from a couple who had been there on the beach for a long time.
9. When first seen, he was sort of on top of her. She was bent into an upside down "U."

Her head was resting in the sand. He was on top, in control. She was submissive.

As we observed, her abdomen would pulse back and forth like my honeybees' abdomens do when sipping nectar. His abdomen occasionally pulsed back and forth in the same way.

They were locked, irrevocably so, either by desire or (as Deana suggested) in the same way dogs lock together at mating, i.e. the male organ swells until the act is completed so that neither can break apart at will.

The wind blew them so that he was rolled onto his stomach. This did not disconcert them, but the new position did affect his posture of "command." He became literally INERT. He could not move. There was an occasional leg movement. In fact, from time to time we thought he may have died. I would touch his leg and he would vaguely respond. In that way we knew he was alive.

She would jerk spasmodically about every 8 minutes. Her abdomen continued to pulse back and forth in this new position. His abdomen pulsed from time to time. I felt that the long time they were locked together---we observed them for about 50 minutes---had something to do with: They had to stay that way until the sperm fertilized all the eggs or they had to stay that way until the sperm was firmly established in her reproductive system so it would not come out when they broke apart.

The tide was coming in. I picked them up with a chunk of sand under them and carried them far enough away from the water.

When I put them down, he regained his original position of hegemony. He came alive again, flexed his wings, stroked her back with his forelegs. Perhaps he needs to be in that position to have life signs. In every other position, he was inert.

A Cosmos in my Kitchen

The last bumblebee mating (which was my first observation) lasted c. 20 minutes from the time I began to observe until they flew away. This mating was still going on after 50 minutes and had been in process when we arrived. So I don't know the length of time of bumblebee mating. Could be from 30 minutes to 2 hours or...?

Fascinating. Locked in a vise-like embrace. Vulnerable. Driven. Committed. One act on the cool, wet sand. Grounded. Grinding out generations.

September 20, 1984

Back from Block Island. Wonderful to be home again with my family!!

Well, since I have written in this journal last July 25, this batch has succumbed. Here's what I think happened to the Bees of '84. To begin with, they were OLD bees. Why?

1. The workers didn't draw wax or gather pollen and nectar as avidly as they should.
2. The queen was old and drone-laying. Not a good queen.
   a. The pattern on brood comb became erratic.
   b. There was more than the usual number of drone cells. That shows that her fertilized eggs were almost gone. Unfertilized eggs become drones in the world of the honeybee. (Fertilized eggs=workers or queens. Unfertilized eggs=drones.) In a hive if the queen dies, a worker bee can start to lay eggs. The impetus for reproduction and for brood seems to be so strong that this does happen. Of course, all of her eggs are unfertilized because she's

never been mated (only the queen gets that privilege). Thus the eggs will develop only into drones. Parthenogenesis (means "virgin birth" in Greek) is what it's called when a worker bee takes on the role of the queen and miraculously can lay eggs.

    c. I observed more than one egg in a cell very often (twin phenomenon). Can this mean Drone-Laying Workers who are dissatisfied with or have lost confidence in the queen?

3. The third reason why I think the hive died and this is primarily THE reason is that the workers weren't healthy, strong or young enough to fight off the DREADED WAX MOTH. The dreaded Wax Moth! The beginning of which I probably saw that time I saw the one-eyed monster in the hive. Also, their larvae may have been/probably were the "twins" I saw in the hexagons?!!

I'm sitting outside on the driveway looking into my opened hive. It is filled with robber bees scavenging fermented and capped honey. The fermented honey is called "honey spoilage."

Throughout the entire hive is white effluvium from which dangles, crawls and rests---the dreaded wax moth larvae in various stages of development. What a disgusting sight!!!

Weak and/or old bees like mine aren't strong enough to fight off this pest and will simply abandon the hive to her and her offspring. I see one gray moth. She's now Queen of the Hive.

I think what happened was:

The bees swarmed or left when I was busy that week in June with the Garden Show at Keeler Tavern. (Won lots of blue ribbons with my God-given flowers. After all, I plant and tend, but He gives the increase. Yet we are given blue ribbons as though we "produced" that perfect bloom!) I couldn't re-hive the clipped queen who was on the ground. She wandered off and died. They returned to the hive, couldn't find her, built a new queen and left. Those bees who remained were queenless. As I've indicated, I think it was, perhaps, a Drone-Laying Worker hive anyway. Workers will take over the queen's function in desperate circumstances. Since they are NOT fertilized, they will be able to lay only drones. (I guess the stinger is a modified ovipositor.) Anyway, the wax moth saw a weak hive and came in. The bees died off or left one by one. The plague took over and here I sit.

These wax moth larvae are interesting to observe, however. There are different stages of its development, too. There are tiny black dots. Bigger black dots. Small, skinny white larvae with a red face and eyes. Bigger, skinny white larvae with a red face and eyes. Fat, healthy, segmented larvae with red face and green-tipped tails. I'm watching these fat ones emerge now from their cotton-like wombs. It is sickening but amazing.

Note inserted later: Re: The "communication" I had with an in-hive bee on 7/25/84: There is a <u>remote</u> possibility that she was trying to "tell" me what was going on in the hive.

I put the honeycombs out for the robber bees. Thousands of honeybees and several hundred scavenger bees took two days to wipe the combs BONE-dry!

What amazed me was something I haven't read about before. Maybe it is too elementary, but it was interesting to me. The bees gobbled up the fermented honey as avidly as the good, capped honey. The alcoholic content of the fermented honey was high. I tasted it! Yet they didn't seem to "stagger" away. By this I mean that they functioned as normally with that honey as with capped honey. Perhaps the alcohol doesn't affect their insect systems as it does the human system?

Here's a great poem I found by Robert Louis Stevenson (of <u>Dr. Jekyll and Mr. Hyde</u> fame). I had to put a good Scottish brogue in my head when reading this. It's the beginning of his poem "Heather Ale."

> "From the bonny bells of heather
> They brewed a drink long-syne,
> Was sweeter far than honey
> Was stronger far than wine.
> They brewed it and they drank it
> And layed in blessed swound
> For days and days together
> In their dwellings underground."

Since I do have a garden composed solely of heaths, HEATHERS and thymes, perhaps my little ones imbibed too deeply of the nectar from these dainty little flowers. Stevenson seems to imply that the honey wine they drank in the poem was made from the nectar of the heather flowers.

My nephew, Mark, Tooie's 2nd boy, helped me pry open the hive with my trusty hive tool. He has been a great help to me. He is very interested in the honeybee and has a talent for natural things.

Still haven't got a final yes or no on <u>Heart of Honeybee.</u>

Re: Robber Bees

I'm sure the phrase among beekeepers, "Once a robber, always a robber," is no more true of apis mellifera than of homo sapiens. I've let bees "rob" my honeycombs three times now and have seen them return to the bone-dry combs, find nothing and then fly over to the nearby heather flower and stick out their nectar tongues. The honeybee will take a free meal, but to her all meals are free. She works as hard at getting the honey from the combs as in getting the nectar from the bloom. Maybe harder. Both are free. Both require work. Who knows, maybe her system is better adapted to storing nectar than storing honey? That's probably true which would mean she would have to work harder and thus wouldn't be tempted to rob once, rob always.

April 22, 1985

I begin my Third Full Year of Beekeeping and Observations.

Yesterday, April 21, 1985 I hived another package of honeybees. That is the 6th package of bees in three years.

I had left the hive in the garage over the winter to allow the deadening cold to kill the remnants of the wax moth larvae, but when I opened the hive this spring, some were still alive!! So much for books that say winter cold will kill the wax moth!

What a massive clean-up job I had!! The wax moths were enmeshed in layers and layers of cotton-like gauze. When I cleaned the combs, I noticed they had eaten huge gouges out of the sides of the combs. I surmise they use some of the wood as food or re-cycle it for the tight cocoons.

Those cocoons were in patches all over the hive. Great swatches of cottony thick material with all stages of larvae developing among them. The outermost larvae were dead, but the innermost were alive and crawling even after the cold winter. The cocoon is so well-made and thick that I believe the young in there could survive almost any temperature. How wonderful! Even though they were an enemy of my honeybees, they are still, as the Scripture says, "wondrously made."

Took Mark and me a total of 12 straight hours to clean, scrape, re-set comb, etc. Mark was a real help. To open the hive, he used a saw to go down between the glass and break the propolis seal so we didn't have to break the glass, as usual. He could be a good bee man. I'm grateful for his help and ingenuity.

The hiving went without incident. Praise God none of us has ever been stung hiving. I used a bee veil. Steve used my nightie. He's really a sight. I have pictures of him in that get-up. In spite of the fact that he is not essentially interested in the bees (he never observes them, etc.), he always helps me whenever I need help with them. He loves me. I love him. Jesse is always right there and is very helpful and unafraid. This year Blake is away at Pitt for his freshman year. He has always been the most interested in the bees. Kathy is in her senior year at Pitt and has really never been here for a hiving. When she is home, she does show interest in the bees, however.

We brought them inside tonight and I am writing this in front of them with their furious buzzing and their unique fragrance filling my ears and nose. I do love honeybees. The Lord knows why, but I do.

I feel I could make some valid contributions to the field if I were correctly placed. That's probably true in many parts of my life. Nonetheless, I have my own thoughts and joys, like everyone does, and have shared them with others around me. Perhaps that is a type of small contribution.

I'm thankful for this batch of bees. I hope they last. I'll enjoy them while they are here and try to learn from them. I'm thankful for my life. I'll enjoy it while it lasts.

April 23, 1985

7:30 in the morning. One observation: They have been laying comb ever since I hived them in the garage 2 days ago. The queen cage was put on the right side. Dummy me. I should have put in on the left side and then the feeder wouldn't be obscured.

When they lay comb, they are packed solid against the foundation. They are not always in a necklace of bees though I have seen that, too. The laying party is often packed side to side, vertically and horizontally, in a cluster. They are so intent, so silent, so absorbed.

The worker bee has 8 wax-producing glands---four on each side of her abdomen. I've seen little scales of wax several times being secreted from a bee that is making wax. Teeny, clear-white and fish-scale shaped. So cute!

Note to me: Remember the fish-scale-like formation the bees assumed when they were caught in a swarm in the storm.

When they are intent on comb building, their bodies almost BECOME the color of the wax. They give the visual impression of a grayish white mass at these times of comb drawing. William Blake's principle, "You become what you behold," seems in operation here.

Beeswax---when I was little everyone used to say in response to a probing question, "None of your beeswax!" That meant none of your business. Why did they use "beeswax" to mean that? Maybe because when honeybees make beeswax, they are so intent and so occupied that they have no time for anything else. "I'm about my own secret business. Don't bother me."

And beeswax is a business for people who raise bees. I bought beeswax candles even before I had my bees. The candles don't drip and smoke. I guess most people would say candles are the most important product from beeswax. But the cosmetic and pharmaceutical industries account for 60% of the beeswax consumption. Beeswax in wrinkle creams, etc. Some cheeses are kept fresh by being coated in beeswax and I capped some homemade red raspberry jam with beeswax instead of paraffin. If wax can help keep a mummy looking vaguely like a human being for 3,000 years, that beeswax can keep my jam fresh for a year or two!

When you think of all the uses for beeswax, it's amazing that ones so tiny can produce so much wax for us! I don't know how much beeswax we use every year, but I do know that it takes 150,000 honeybees to make just 2 pounds of it!

10:30 P.M. Put in a new bottle of sugar water. Looked 5 minutes later. The darned thing had all leaked out onto the floor of the hive!! I've had SO much trouble with that-----BOTTLE!!! The entire design of this hive leaves a few things to be desired:

1. Design of the way bottle affixes to hive and the angle to maintain the vacuum.
2. The hive is very prone to condensation.

BUT "Praise The Lord ANYHOW," as the saying goes.

They have their nectar tongues out, rear ends pulsing away and they are eating up the sugar water. They hang upside down from the bottom of the brood comb to stay out of the sticky goo.

Actually, this is a good opportunity to study the Nectar Tongue.

The nectar tongue seems to work thusly: It appears to be one long red shaft. It is inserted into the sugar water. The tongue shaft opens into two parts like a "V." From the center of the "V" a feathery pink shaft appears which is longer than the central "V." That portion seems to probe and aid the suction. I notice little feathery wisps that are also there and are used to go around the debris on the bottom of the hive. Those feathery things (will have to look up what they are called---did---"labial palpus") whoosh the sugar water into the "V" and thence down the alimentary canal into the nectar crop in their abdomen. This is, also, I'm sure how they gather and store nectar.

1. As each bee sucks, her antennae go up and down.
2. The longer feathery tongue in the middle darts in and out in a probing manner.
3. I notice that when the feathery tongue is inserted into the sugar water, the probing action of this part of the tongue produces a round circle in the liquid.
4. Four fluid ounces of liquid leaked onto the hive floor. There were about 50 bees working full-time to suck it up. It took them 33 minutes to clean up 4 oz. of sugar water. And they cleaned the hive floor bone-dry!

5. There were dead bees in the bottom of the hive who were soaked by the sugar water. The bees used their nectar tongues to lick them bone-dry, too.

As that emergency was being handled by the 50 bees, the rest of the hive continued laying wax, delivering messages, etc. They were, to the human eye, oblivious to the emergency. Or is it that they had SO MUCH CONFIDENCE in the part of the body that tends to those things that they had utter faith it would be handled?

Oh, that we in the Body of Christ could emulate the bees' ability to trust each to do his/her job. But we all must be as trust-WORTHY as these creatures He has made.

April 24, 1985

Full 4 oz. bottle put in at 4:15 this afternoon. 50 degrees. They haven't been out all day. Are laying brood comb and honeycomb.

April 27, 1985

Midnight. Tonight we set the clocks forward. "Spring forward, Fall back."

Tomorrow, April 28th, is my father's birthday. The bees' 6 week life cycle must seem as long to them as his threescore and ten seemed to him. He died March 12, 1979. It seems like he is just away, not dead, and that I'll see him walk through the kitchen door again. Well, praise God, I will see him again.

This whole side of the brood comb is almost ready for eggs. It's taken them only 5 days to do this side and I'm sure they've done the other five sides, too.

Pollen is packed in randomly on the back of this comb. I can see its golden color through the wax. There is typically none on this side. They hide it. But this side does glisten with nectar. As usual, it is in the hexagons on TOP of the brood comb.

This side is now packed with bees since I've opened it. I love them. They love the light and every batch of my bees have preferred the open side of the hive to the closed one even though the booklet said they may resent observation. (The one batch where the bee gate was put on backwards was the exception.) They don't resent the intrusion, I find. They thrive on observation in this kitchen. They know they're loved.

April 29, 1985

The hive glass has condensation on it. I open the top of the hive to let in more air. That helps a little. It is NOT muggy outside. We are in a prolonged period of drought. So the condensation is caused by the design of the hive, the press of bees and their activity.

A few apiculturists have suggested that the water is caused by the evaporation of water in the nectar. That argument holds no water (ha, ha) because the condensation was on the glass in the first 24 hours after hiving. There was no nectar in the hive---only comb foundation. They didn't leave the hive to get nectar for 48 hrs. after hiving. Having 10,000 bees in a 3 comb hive without nectar produces water, condensation---somehow.

Sandra Sweeny Silver

April 30, 1985

Re: Drones Begging For Nectar From The Workers

The drones are VERY aggressive when they want food from the nectar crop of the female. They are at this time of night (midnight) usually quiet and resting on the top half of the brood comb. But if hunger pangs begin, he will leave his resting place and go in circles, willy-nilly, petitioning each passing worker for food. Most workers either have none to give him or are busy with something else and don't respond to his overtures. But the drone will always find a willing one. Often a worker approaches him and, unsolicited, will give him nectar from her crop. It usually takes 2-3 workers to satisfy his hunger and then he will go back to his resting place. Nectar contains up to 80% sugar. Guess those drones needed a nighttime carbohydrate fix!

As he extends his nectar tongue, his and her mouths intertwine. (This is, perhaps, a form of sexual contact and satisfaction for the female worker?) He aggressively feeds, pawing her head and forelegs with his forelegs. He seems to feed until SHE terminates and wanders away. He then frantically searches for another worker.

I don't find the drone as pitiable and impotent as the books and other observers find him. I see him, them, as kings of the hive who are tolerated and coddled like the queens are.

The rulers of the hive, in truth, are the workers. But it is all so symbiotic. Each needs, cannot exist without, has a meaningless life without---the other.

May 10, 1985

Blake got home from his 1st year at Pitt tonight---Hallelujah! Had a huge pot of vegetable soup waiting for him. He loves my Mother's Vegetable Soup!

Kathy got home from her 4th year at Pitt on the 6th---Hallelujah! I made her favorite dinner. Pot roast with mashed potatoes and gravy, petit pois and YORKSHIRE PUDDING. She loves those high and fluffy treats with lots of good, brown, homemade gravy filling the nooks and crannies. We all do!

My three lambs of God are here again under the same roof!! I feel complete when they are all here.

> "Little boy blue,
> Come blow your horn.
> The sheep's in the meadow.
> The cow's in the corn."

A simple nursery rhyme (if the deep truths of the songs of children can ever be called "simple") to illustrate the fact of Topsy-Turvy. We must be prepared for the unexpected, the "that's-not-supposed-to-happen-ness" of life. Tonight the unexpected happened.

Blake and I were observing after he arrived home. Blake said, "Is that the queen?"

There her highness was and is now up on the honeycomb!! She's loose in the upper hive like the cow in the corn. She's not supposed to be there like the cow is not supposed to be loose in the cornfields. The metal Queen Excluder is to keep her down on the brood comb, so she will not lay brood in the honeycombs.

When I hived her, I observed she was very small in comparison to my other queens. The drones in this hive are as big as she is and their abdomens are as wide. They are all Carniolan bees with an Italian queen.

Carniolan bees are similar to Italian bees in appearance. Some beekeepers like them better than the Italian bees because they have longer nectar tongues and that is reputed to give them an edge in nectar storage. They are, also, gentle like my previous Italian bees. Probably the Italian bees and the Carniolan bees (I think originally from Slovenia) are the two species most used by beekeepers.

Back to the sheep's in the meadow and the cow's in the corn. My queen is not where she is supposed to be. The queen excluder with its metal bars has one weak spot anyway. She either got through that spot or she is so small that she can go through the bars. Now she will be able to lay all over the hive! That's a no-no. She's only supposed to lay on the bottom brood combs in man-made hives.

My other queens were too fat to get between those little metal bars.

There's nothing I can do about this, so I'll do what I always do when something goes wrong: I'll observe the situation as I find it. That's why I have this hive anyway. I don't want to manipulate them for honey, wax, pollen or propolis. I want to observe them in their natural state. In a wild hive she lays all over the place, so this will be like a wild hive.

May 30, 1985

The bees are swarming!!

A Cosmos in my Kitchen

It's 1:00 in the afternoon. Sunny. 70 degrees.

Can't find the little clipped queen. They're on the same evergreen tree as all the rest of the swarms---but on a different branch of the tree. They all seem to pick the SAME tree, but a DIFFERENT branch. This swarm is hanging in a long c. 24" beard.

Note: Saw a picture in a newspaper of the "World's Biggest Bee Beard." The man is in the Guinness Book of World Records as having "worn" a beard composed of 35,000 bees weighing 10 lbs. The article says "he created (the beard) by affixing a queen bee in a tiny cage to his neck, then having the worker bees, which were drawn to the queen, released." He's Jim Johnson of Terra Alta, W. Va. Most of the beards I've seen do not cover the entire face of the person. On Jim's face the only part NOT covered is his nose and a part of his mouth directly under the nose.

Have looked all over the ground for the queen. Can't find her. It's 1:45. I'll wait and see if they come back or if they have killed her and swarmed with a new queen.

3:30 afternoon. They're gone. So they had a viable queen and killed the little one. Maybe she squeezed through the queen excluder to get away from them!

Hope the ones who are left have a queen and will persevere.

The hive looks tidy, busy, half-full.

June 14, 1985

The queens are piping, buzzing. Something's up. Yeee. Yeee. That piping sound is maybe a G sharp or an A natural? Tonight the concert is a two second whirr (of the wings?) and then 4 toots (4 blows of the pipe).

Have been boning up on mythology by rereading Bullfinch and Larousse. Found out that the demi-god Pan was the patron of beekeepers! Maybe they were making mead way back then! Ancient statues of Pan show him as a half-man, half-goat creature with a reed pipe to his mouth or in his hand. Because I believe every myth, legend has a kernel of truth, I've always thought these satyr creatures meant: man has a sane, divine nature AND an irrational, bestial nature. Pan is associated with the god of wine Bacchus. Pan blew his reed pipe and many Romans responded to a spring rite called the Bacchanalia. This unholy rite was a drunken, drug-filled, sex-drenched orgy that lasted for days. The Roman Senate outlawed the rite, but you can't stamp out man's fallen nature. In 1967 when I covered the Hippie "Summer of Love" in the Haight Ashbury for a Pgh. newspaper, I reflected that it was just a three month long Bacchanalia. It's interesting that Pan was called The Great God by those who worshipped his hedonistic, licentious ways. He was like the Pied Piper of Hamlin. The music he played on his pan flute was so alluring that many left the possibility of a sane life and followed the sweet music of destruction. For the music must have been deceptively sweet, right? The Bible says that in order to entice people, Satan disguises himself as an angel of light. (Again Shakespeare's Appearance vs. Reality theme.) Pan's pipe is still around and called "the panpipe." It's a wind instrument consisting of pipes of varying length held together by metal or clay. The ones I've seen on ancient statues have between 2 to 8 reeds. I know that some Peruvian pipes have many more reeds. I LOVE the music made by Peruvian pipes. Through hearing that contemporary Peruvian music, I can understand how alluring Pan's piping must have been. I think of the "high Yeee sound" of my queens.

It HAS to be a "wind instrument" sound. It's like a pipe and not a reverberation. I think the reason Pan was patron of beekeepers was because he was a rural deity who lived in the mountains and countryside where bees are most abundant. Also, there is that honey and mead alcohol connection. Pan lives on with his pipes and in our poetry. The nature and meaning of Pan has attracted many poets. The pagan William Wordsworth loved all the pagan deities and longed to live in that world:

> "...Great God, I'd rather be
> A Pagan, suckled in a creed outworn,
> So might I, standing on this pleasant lea,
> Have glimpses that would make me less forlorn:
> Have sight of Proteus rising from the sea,
> And hear old Triton blow his wreathed horn."

The Christian poet Elizabeth Barrett Browning wrote a <u>long</u> poem declaring Pan dead. Here's several snatches:

> "Earth outgrows the mythic fancies
> Sung beside her in her youth;
> And those debonaire romances
> Sound but dull beside the truth.
> Phoebus' chariot course is run!
> Look up, poets, to the sun!
> Pan, Pan is dead.
> And that dismal cry rose slowly
> And sank slowly through the air,
> Full of spirit's melancholy
> And eternity's despair:
> And they heard the words it said---
> 'Pan is dead! great Pan is dead!
> Pan, Pan is dead!'"

But Pan is not totally dead. The word "panic" survives as a sudden, overwhelming fear that can spread quickly through your whole body or through the body politic. The "pan" in that word surely connotes the uncontrollable behavior at the Bacchanalias. And "satire" comes from "satyr," the half-man, half-beast exemplified by Pan. Looked up satire: "the use of irony, sarcasm or ridicule in exposing vice or folly." Would I be sarcastic in saying: "Yea, sure, of course the great god Pan is dead?"

And so to bed.

June 15, 1985

The bees swarmed again!! The swarm is small---half the size of the May 30th swarm.

Last night I heard the piping. Yeee, Yeee, Yeee, Yeee. I'm SURE the queen was communicating to her troops that she wanted to get out of here. DEFINITELY THE PIPING PRESAGES AN IMMINENT SWARMI! (Among other things, I'm sure.)

Please--I hope they left another gyne cell so that I'll have a queen.

11:00 in the morning. It's sunny. 68 degrees in the shade. 75 degrees in the sun. They have been hanging on a branch of that same yew tree for an hour. The swarm is tight and all are hunkered in except the few who riot around the swarm guarding it.

Beekeepers often contact the police to let them know that they will come and get any swarm of honeybees. Normally swarms are pretty low to the ground and are easily captured by an experienced beekeeper. They approach the swarm and try to determine if there are mites or other insects that would make it an undesirable swarm for him or her. If I had called a beekeeper to get any of my swarms, he would have been able to capture the swarm easily.

They have always gone on a branch of that yew tree by my kitchen window. There they hang about 5-6 feet off the ground. The beekeeper usually has a box. He places the box directly under the swarm and forcefully shakes the branch downward. He hopefully gets most of the swarm into the box. The rest he sweeps in with his gloved hands. Behold, he has a new hive of bees for nothing. I think of my friend Larry from West Virginia. He crawled 70 ft. up a tree with a plastic garbage bag and smacked and brushed a swarm into the bag and then climbed down again. He was only 12! I've seen pictures of swarms. One of them had settled under one of the seats of a picnic table. But that is unusual. The swarms are usually in trees.

June 17, 1985

At 4:30 today the bees finally left and went to their new home wherever that is. They hung there in swarm formation for two full days. It rained torrentially the night of the 15th when they swarmed and last night. With all the rain the scouts probably couldn't get out to find another home.

During the Bible Study last night Mark and I were worried about the bees and the lack of food for so long---even though they engorge with honey before leaving. So he and I made up some sugar water and squirted the swarm with it. The squirter wouldn't work right, so we taped a piece of honeycomb 2" from the swarm. It rained and the comb fell down. They seemed to survive well enough to leave today.

No matter how close I got to the swarm---and I almost touched the entire swarm, cradling them with my palms---not one as much as buzzed. Emboldened, I STROKED the swarm. Mark got as close as I. They wouldn't dare sting him because he was helping me and them so much!

I read a bee article about swarms once and the author said you could plunge your hand into the middle of a swarm of 26 lbs. of bees (tens of thousands!) and they wouldn't sting you. (Engorged with honey, they are placid. "Swarm bees are sweet bees.") The writer didn't say that he HAD DONE this so I think he was hypothesizing. Anyway, I wouldn't do that. It would disturb them too much and, well, you know, I just wouldn't do that! I may be foolhardy, but I'm not foolish.

July 14, 1985

Bastille Day.

Even though two swarms have spun off---"Allons, enfants"---the hive is still viable, active, carrying in pollen, storing nectar. I think there is a queen.

I must say, these Carniolan bees are the best I've had. Industrious.

Of course, the honeybee has since ancient times been a SYMBOL of hard work. I love numismatics, the study and collection of coins. I've collected coins sporadically for over twenty years. In one of my books there are pictures of ancient coins. One is from Crete and has a goat on one side and a bee on the other. Another one (400 B.C.-148 B.C.) has the goddess Persephone, the reluctant queen of the underworld, on one side and a honeybee on the other side. Definitely the ancients weren't worshipping bees. They were using them on their coins as symbols of hard work and productivity.

I love to see them out in the gardens. Right now in the back garden the bee balm, cimicifuga and liatris are blooming---lovely spikes of white and pink and fuzzy faces of red/pink. Solitary and honeybees are on them gathering nectar and pollen.

September 26, 1985

Since 10:00 this morning, Hurricane Gloria has been pounding Ridgefield. We've had 40-50 mph wind gusts. It's the first hurricane we've had since we came to Connecticut in 1972. The electricity went out at 9:07 this morning. Branches and debris have been flying around outside. We have stayed in our kitchen. Steve and Jesse are building a cage for our two ferrets right here on the kitchen floor. One ferret is Jesse's and one ferret is Blake's.

Blake took his ferret to Pitt with him, but the ferret chewed through his roommate's water mattress. Guess who paid the $500. for another water mattress? So back home the ferret came. The boys really wanted these ferrets, but guess who takes care of them and cleans up after them?

I'm sure we won't have damage from Gloria. This old Victorian house has seen Gloria and worse in its 100 years.

Most of the damage so far is water damage. There are 2"-3" of water in our basement which is a first for our ten years in this house.

The windowsill where the hive rests is flooded. Right at this moment, we are in the eye of the hurricane---2:15 in the afternoon. There is No rain. The SUN is popping in and out and it is balmy! There are times of respite and joy even in the midst of the storms of life.

Believe it or not, the workers are leaving the hive to forage!! They obviously don't KNOW we are still to get the tail end of Gloria!

We just heard over the battery radio that wind gusts in Ct. have reached 92 mph. Gloria is really mad!

October 14, 1985

Most of the bees are dead or dying, but I believe a nice cluster will survive. There are no stores on this side of the hive, but I assume there are on the inside.

These Carniolans have been the best strain I've had. They are prolific, hardy and industrious.

The old brown cells, shiny with propolis, stand empty and neat. Hopefully in the spring a new batch will be hatched.

Thank you, Lord, for my bees and their lessons to me.

October 22, 1985

Last night there was another 28 degree frost. Today hundreds of bees are dead in the hive and on the bottom of the brood comb. Fifty are dead in the lower left hand corner of the honeycomb.

It is a magnificent, sunny, crisp, 60 degrees-in-the-sun Fall day, but they are hive-pent with their dead sisters. Several dead brown leaves are in the windowsill outside the bee gate. Everything is bending and dying as the sun journeys across the sky to wintertime.

I love, love Kit Williams' book Masquerade! And his second Book Of No Name. No Name is about the honeybee and the seasons. Both books have magnificent drawings! Glorious text! Requiring thought and inspiration! My kind of guy!

In the honeybee Book Of No Name (the reader has to figure out the name of the book!), Williams depicts the seasons. He draws Summer as a yawning little fat boy with a bee in his mouth!! Winter is a very sophisticated middle-aged thin man dressed like a funky courtier. He has the Four Seasons sort of gobbling up each other. What a brilliant genius is Williams!! I hope he writes and illustrates a lot of books!!

And it is Fall here---everything is falling. ("Fall" comes from the Old English word "feallan" meaning "fallen." Our season Fall is so named because everything is "falling" away.) The dead in and out of the hive have served their purposes. Now their discarded shells will be used to replenish the earth. Those hollow shells that contained the living. Our body, our shell, contains us, the real us no one can see.

Note inserted later: I have a shell collection at the entrance to my home. Steve's mother, Natalie, gave me her collection when they moved to Ormond Beach, Florida. I don't know a lot about shells but they are beautiful, unusual, amazing. And shells once had living creatures inside of them. I have a big, pearl-colored chambered nautilus shell. When I looked it up in the shell book, it says it's related to the octopus! It has a shell that is divided into what they call "chambers" or I'd call segments. The more I read about it, they should be called "chambers" because as the nautilus grows bigger, it seals off one segment and secretes a new, bigger "room." As the chambers are created, an elegant spiral-shaped shell is produced. This spiral is called a "logarithmic spiral." In this type of spiral the chambers increase in length and breadth at a steady but unchanging ratio. I can't get into the math of it because I'm not good at that. But I can grasp the concept. Descarte in 1638 made an equation for this spiral of the nautilus shell. The essence of his math is, I think, size increases---shape does not. Other logarithmic spirals are: the eye of a hurricane and the spirals spinning around it. Spiral nebulae in space have this shape and even babies developing in the womb do! So this seems to be a God-shape. Some mathematicians link this spiral to the Golden Ratio. I'm definitely NOT qualified to think about that! But there was Jacob Bernoulli (1654-1705) who was decidedly qualified. He studied the nautilus shape so profoundly and he was so impressed by its mathematical and mystical proportions that he named it "spira mirabilis" (the "miraculous spiral").

Amazingly, this spiral is, also, linked to Fibonacci's numbers. I love these scientific types! The miraculous spiral is the GEOMETRIC PATTERN corresponding to the NUMBERS of Fibonnaci. Way, way beyond me when I look at all the equations! But I sure can glimpse the interconnectedness of ALL creation through this little exercise. Bernoulli had engraved on his tombstone: "Eadem mutata resurgo." He understood like I do that death is just another chamber we enter. The Latin phrase means: "I shall arise the same, though changed."

I'm making sugar water for them. They had sealed over the hole in the bottle with propolis. I'll start supplementing this afternoon.

A Cosmos in my Kitchen

October 23, 1985

I can't believe this! Kit Williams is designing a clock to be placed in the middle of a shopping center in England! Wow! It will mystify and amuse forever.

Note inserted later: The name of the clock is the Wishing Fish Clock. It's in the Regents Arcade shopping center in Cheltenham, Gloucestershire, England. It's 45 feet tall. But... (Sandy, you love this stuff. Williams and you are kindred spirits.) He's got a duck/goose laying golden eggs continually. (The goose that lays the golden eggs of Precious Time. Try to steal him, Jack. Use that currency wisely.) A fish at the base of the clock blows bubbles every hour. (Make your wish for more Time. The fish/Pisces is from the watery depths of the Spirit that is eternal/outside of Time as are the evanescent bubbles.) The clock has mice (that nibble away at Time). They are trying always to evade a big snake that sits on top of the clock. With the snake, the Uroboros, Williams has really tied it all together. The Uroboros is a snake/dragon that swallows its own tail. It is in a constant state of creation and is a symbol of the Wheel of TIME. PLUS, the Uroboros ties in to so many ancient symbols and equations that attempt to unite in ONE all knowledge and lore. E.g. The archeometre. The archeometre looks like the Aztec calendar and is an ancient correspondence figure. It purports to contain all color, music, zodiacal signs and other arcane information. The Uroborus is, also, the "12 Around 1" symbol personified by Jesus and His Twelve Apostles, The Twelve Tribes of Israel and the Twelve Signs of the Zodiac, etc. There are other ancient and esoteric correspondences. Many came from Gnostics and alchemists. I've studied these and find them interesting but essentially ONLY CEREBRAL. I love cerebral, but I live in a Newtonian pragmatic world! I can't help but say that the Uroborus, also, is connected to the Golden Ratio of Fibonacci and the Greeks. AND the spiral nautilus is part of the Golden Ratio and the flowers of my bees are part of Fibonacci's numbers and Fibonacci's numbers are part of Sacred Geometry as is the Uroboros and it ALL is part of the "GLORY OF GOD IS TO CONCEAL A THING." Williams seems to have tied up so much glory in a Mall Clock in the middle of a pedestrian shopping center. God. Bless him.

There is a 3" pile of dead bees on the lower left hand side of the brood comb. It is about 6" long. There's another pile on the bottom of the honeycomb about 1" high and 6" long.

They took one full bottle of sugar water yesterday in 8 hours. Took almost one full bottle in 8 hrs. today.

November 11, 1985

On days that permitted, the bees have removed all the dead bees (and some dead flies) from the hive.

No stores on this side. They are taking one and a half bottles of sugar water every day.

They have made their winter cluster on the lower right hand corner of the brood comb near the feeding bottle. In other years the cluster has been smack in the center of the brood comb.

Today is and tomorrow is supposed to be a late touch of Indian Summer. 60-70 degrees. Now at 7:00 in the evening the bees are lolling and roaming all over the brood and honeycombs. (Love the saying, "Today is the Tomorrow you worried about Yesterday.")

A Cosmos in my Kitchen

All the leaves have fallen off all the trees. In spite of this lovely weather, we had our first snowfall two days ago. Early winter has moved in with his bare bones and chill breath.

BUT my herbs are still flourishing. This afternoon I made Mother, Corinna (the Swedish au pair who has been here for 6 months and leaves to go back to Sweden the day after Thanksgiving) I made the three of us hot cups of fresh herb broth. Chopped parsley, sage, rosemary, thyme, lemon balm, chives, oregano and shallot shoots in chicken broth. Oh, it is so good!! Every week or two, I make us an afternoon cup of hot water with sage leaves and honey. Sage has such a haunting flavor. I love it. There's an old Persian proverb: "He who has sage in his garden should never die." I use these herbs a lot in salads, infusions, soups. I'm not into them because of their health value but because I LIKE them. Everyone is so health-conscious nowadays. Can't eat this. Must eat that. Must exercise. Don't do. Do do. I hate all this emphasis on the body, the body. People should be more concerned with their minds and spirits than with their bodies. At least, AS CONCERNED! No one ever says their mind is flabby or their spirit needs tended. It's always that which WILL PASS AWAY VS. that which WILL ENDURE FOREVER.

I'll be so sorry to see Corinna go. God has sent us these wonderful jewels from Scandanavia: Lena, Gunilla, Ainsi, Corinna. All four of them have given their lives to Christ while here. Blake led Corinna to the Lord. I consider them as seeds of Life to be sown over there for future generations to rise up and call Him blessed.

Most of the flowers are dead, even the lowly, reliable marigolds. Sunday in preparing the Bible Study bouquet I managed to forage some willowy juniper boughs with the ice blue berries and a few frosty mums. Plus I filled a

Mason jar with several snapdragons, two marguerites, two red clovers, several geranium lancastriense. They were lovely. I have MILES of tenacious bittersweet hanging over the tops of the tall evergreens out near the pool. I pulled some down and wove them into two grapevine wreaths I had. They look nice.

I read they found pollen grains in the grave of a little prehistoric girl. How touching. I can imagine her loved ones gathering all the flowers they could find in order to place them lovingly on her young body before consigning this little one to the earth. When we lived in Edinburgh, I read that in a bronze-age burial cairn they had found pollen from the meadowsweet flower. And pollen from one of my favorite flowers, the anemone, has been found under and over the bones at a Stone Age burial site. We bury our dead with flowers. If someone dies, I don't send flowers to the funeral home. I always send them to the family. Seems I am out of synch because all these pollen grains are always found where the dead are. I love the fact that when humans die, their essence or soul goes on forever. And the flowers die, but the essence of the flower, the pollen, the seed of Nature, goes on and on. A teeny part of the Bible Study bouquet lives on.

April 27, 1986

This April starts my fourth year of beekeeping.

The good Carniolan bees died a natural death of dwindling over the hard winter plus there was residue of the wax moth problem.

I was thinking that the silk worm if left alone would develop into a moth, too. The wax moth invasion in my combs gradually spun all that strong white material for their cocoons all over my hive and destroyed my bees. But the silk worm is never allowed to develop into a grayish-white moth. About 4,000 years ago, the Chinese learned that the larva from that particular moth makes its cocoon by spinning up to 1,200 delicate, silken threads. Someone, somewhere had the idea to take the threads and make them into material. The material and where it came from was so guarded by the ancient Chinese that anyone who broke the secret or smuggled out silk worms was killed. The Chinese "farmed" these insects by placing the worms on their favorite food, mulberry leaves. They ate and ate for 6 weeks. Then the larvae started to spin the silk threads and cocoon themselves. The "farmers" killed the larvae in the cocoons by dropping them in warm water. Then began the painstaking and delicate task of unwinding the silk threads cocoon by cocoon. 10 or 12 of these threads were combined into a single thread of raw silk and wound up on a skein. From there the thread was woven into the luminous material that made China famous in the west. Certainly the material that the wax moth larvae exuded all over my hive would not be described as "luminous" or "silken." Nor can I ever imagine that dense, cottony material causing such a sensation that an entire highway would be named for it (e.g. the Silk Road). But it was sturdy and strong and resistant to eradication. That's one thing that the delicate, silk-spinning cocoon of the Bombyx mori moth doesn't have going for it.

Today I installed a new cage of bees.

Opened the queen cage. She was dead. Had to interrupt the hiving, drive down to Wilton, get a new queen from Ed.

The hiving went well after that. Because the bees needed MORE time to get used to the scent of the new queen, I did not puncture the candy plug to facilitate their gnawing as I usually do. Normally, the bees have been in transit for several days with the queen cage hanging there amongst them. But since this queen was dead, I gave this new queen more time to be accepted by them.

Nonetheless, it took them the SAME AMOUNT OF TIME to gnaw through the candy plug and release her as it did the other batches where I had punctured the plug with a nail. It takes them about 18-20 hours to release the queen.

Now they are happy and busy at work drawing new brood comb. I had to tear out and replace the old brood comb because there was a little of the wax moth cotton-like excretions.

When I took apart the brood combs, I found a lot of capped honey on the inside of the hive. I strained it. It was pure WHITE!! I'm sure it was gathered from the red raspberry bushes growing near the tennis court. White honey is very rare. I have LOADS of the English variety of red raspberries---thousands of tiny blossoms---that give us a big bowl of raspberries every day for a month!! The raspberries, also, gave my bees a big comb of nectar. It is of excellent taste and consistency. A real hidden treasure---WHITE GOLD!

Note to me: Remember, Sandy, when you wrote you would love to taste white honey? (2/12/83)

May 19, 1986

These particular bees really must love honey! ALL of the honeycomb is already filled with nectar and two thirds of it is already capped. Plus, the whole top 3" of the brood comb is capped honey and half of the remaining brood comb is either capped honey or glistening with ripening nectar! All that work in only 22 days!

My other batches of bees put no more than a fourth of the brood comb in capped or nectar cells.

Also, there is no developing brood on this side of the glass. Perhaps they killed the queen I had put in? That's one explanation for so much honey, no larvae.

Are they making a new queen? Can they? Only parthenogenesis under certain circumstances?

How I fret. There are, however, cells that are packed with pollen so there probably is a queen.

Just learned that chocolate-coated pollen candies are sold in some health food stores! Now that's a sugar-coated pill! Pack in the protein! Get your chocolate-covered pollen bars!

May 21, 1986

No brood on this side, but there's pollen. I assume they wouldn't pack pollen sans larvae.

I don't know if I've written about this before, but I've observed this many times. Re: Color of Wax

There is a difference in the COLOR of wax used to cap the honeycomb cells on TOP of the hive and the wax used to cap the honey or brood cells on the BOTTOM of the hive. Usually the honeycomb cappings are WHITE OR GOLDEN. The brood comb cappings are BROWNISH.

Now, perhaps I should mention different colors of wax. In this particular hive, the BROOD comb honey is capped with SNOW WHITE WAX.

The HONEYcomb honey is capped with HONEY-COLORED WAX.

The wax hexagons themselves start out pure WHITE. Use of them plus propolis smearings turn the hexagons BROWN.

I replaced the brood combs on this hive with pure white foundation. They drew white wax. The cappings on these are pure WHITE.

I left the drawn honeycombs as they were. The cappings on them are GOLDEN.

Why would the cappings on one be white and on the other golden?

September 7, 1986

This has been a great batch of bees. I have been extremely busy writing books and observing all kinds of things other than bees this summer: Early American and Japanese Quilts, Oriental Rug Patterns, Illuminated Manuscripts, especially the Spanish <u>Beatus</u>. (Went to the Morgan Library and saw some of it!) But the bees have been thriving. I do check up on them every day, but I haven't had time to astutely observe and record.

They are now a part of my life. The initial excitement and flurry is over and they are now incorporated into the overall structure and meaning of my life. I would miss them as I miss all loved ones far away or dead. I love the FACT of so much life, so much industry, so much meaning going on at all times in the left hand side of my kitchen.

January 30, 1987

As can be seen, I haven't written in this journal in 4 months. But I observe them and love them. I suppose if I got a snake or a chimpanzee or---but those are not good analogies. The honeybee hive is a world all its own. Self-contained. Analogies will fail because nothing really beats having a city of 10,000 beings birthing and going about their business and warring and dying right there in your own kitchen window!

36 degrees. Sunny. Ground covered with 18" of packed powder! Ice. Icicles. Winter limns everything.

Because of the sun, several bees are flying in and out of the hive---taking out feces, frolicking. One bee is frozen on a fallen icicle below the hive. Another bee is immobile in the snow under the hydrangea bush still sporting her dried white blossom plates. But just this moment, 10:21 A.M., a little bee came right into the bee gate. She had rounded the bend near the window and flew right in! It's 36 degrees at the bee gate and sunny. Who said bees can't navigate in cold weather?!

Now three bees are circling around outside the hive window. Three bees are walking gingerly over the icy snow in the windowsill outside the bee gate. Now at least 10 bees are either flying in circles outside the window or walking on the crusty snow in the sill. I see two more tiny bee bodies in the snow---bits of black in a basket of white. I watched these two round the corner and they either fell or landed in the snow. I'm sure they are frozen now.

One bee had been trying to get over the snow in the sill to the bee gate for about 15 minutes. She didn't make it and has become a hole in the snow.

The sun has just gone behind a large dark cloud. No more bees are venturing out. That was an expensive trip for some!

An interesting thing: As the appointed or adventurous bees were flying out and either dying or returning, the excitement level in the hive approached swarm energy. The whole hive became exhilarated that some were actually leaving the winter hive and going OUT IN THE SUN!

Some bees did survive this Polar Bear Club activity and returned in tact to the hive. The books say that they cannot survive in 36 degree temp. Never say never with the honeybee!

This batch has definitely been my best. It is almost February and the inside of the hive looks like June---clean, crisp, moisture-free. The bees that are still living are active and pristine in color and texture.

I hate to say it, but I have done the LEAST with these bees. All summer I was frantically typing and writing that cookbook with Maria B. God only knows if it will EVER be published. It is a bear of 600 pages!

In the fall I noticed that the honeycomb on this side was empty. Most of the other batches went into the winter with this side full and capped. I thought maybe they wouldn't have enough stores to see them through the winter. I've given them only three or four bottles of sugar water all winter versus almost continual sugar water to others.

A Cosmos in my Kitchen

This group has survived and thrived by letting Nature take its Course. Makes me feel irrelevant, and, of course, I am.

They, like the others, have propolised over all of the four air vents in the hive. They are tucked in totally on their own and doing well.

In mid-January I could tell the queen was beginning to stir, to lay. Hive activity went overnight from a relatively loose tight cluster who were at ease and meandering to---no cluster and bees all over the comb, excitement level high, purposeful movements to and fro. It had nothing to do with the outside temperatures (which can occasionally alter the behavior of the winter cluster). The thermometer hung at 20-38 degrees during the day; 0-10 degrees at night. I knew she was gearing up again. They had "meaning" in their lives once more. New generations were being hatched!

In December a radio talk show in Greenwich had me on to talk about honeybees. They hooked me up with them and their callers while I remained in our kitchen. The host originally said I would have a 10 minute segment, but there was so much response to my information that I was on one and a half hours.

The host was flabbergasted that people could be so interested in "bees." But I know that ANYTHING can spark interest in people IF the INTERESTING things about the subject are conveyed. Rather than let him ask questions about the honeybee (he didn't know enough to ask the right questions), I told him to just let me talk about 10 minutes on the air about them. I told a lot of the facts about how many flowers visited per day per bee, all the really fascinating and amazing facts about apis mellifera. That sparked enough callers for the rest of the show.

I sure would love it if the Lord would allow me to publish my Honeybee Illuminated Ms.

Right now I'm feverishly at work on a book, <u>Abortion: A Biblical Consideration.</u>

We Christians have sat on the bleachers while the Devil has played his game with the lives of the babies. We have watched in horror as 1.6 million babies are aborted each year JUST here in America. It is time we all stood up, repented our spectatorship and became the Opposing Team. The book is going to be based on all the Bible quotes which speak to and cry out for the sanctity of womb life. I've gone from Genesis to the Revelation and have hundreds of verses that I'm organizing into three main categories. I'm doing essays on each category.

Plus I have started picketing an abortuary in Danbury on Saturdays. I am quite bold and have found the cars that pass by either give me the finger or a thumbs up---which about sums up the conflicting views in America in 1987.

Note inserted much later: Published the book in 2002. Chuck Colson gave me a nice quote to put on the front cover: "I've never seen more comprehensive research on the abortion question...a tremendous resource for people." Charles Colson

Chuck's letter to Bob B. who sent my ms. to him goes on to say: "Please do thank Sandra for me. I believe this is the decisive issue of our age….the abortion issue may well cause our social contract to be unwrapped. Christians need to be deeply concerned and involved. So we're grateful to Sandra for this massive effort."

February 14, 1987

Happy Valentine's Day, my sweets! Cupid used to dip his arrows in honey before he aimed them at lovers. I love you and the time I spend with you. I'm "smitten," as they used to say. Smote with an arrow. A honey-dipped one.

10:50 A.M. 42 degrees. Sunny. It was minus 5 degrees last night. Has hugged the single digits for days now. But on this 42 degrees day of respite my bees are literally risking their lives and going outside!

There are dead ones in small circle tombs in the snow outside the hive. The sun is directly on the bee gate and this seems to inspire some of them with false confidence. Inside they are going crazy with activity.

Some seem to become disoriented outside in the cold. They try to find the bee gate and will hit 12" to 14" away from the hole. They become too cold and fall to their deaths.

I have given them one and a half bottles of sugar water in the last week so don't feel they are impelled by nectar need to venture out. (There IS no nectar.) I think they are just hive-pent and will risk known death for a free fly in the warm winter sun. Is there a corresponding need in us?

The hive side of the house is stained brown by the bees. There's a 1ft. to 6ft. stain that goes up three stories! Is the brown stain feces, propolis, something else, all the above?

But this hive is without question the healthiest one I've had on a mid-February morning.

May 4, 1987

They are parading a torn-out queen larva and a dead, partially developed worker bee. The queen was undoubtedly killed by the present queen, but what about the worker larva?

May 18, 1987

10:00 A.M. 85 degrees. The first swarm of the season!!

They have swarmed on the exact same evergreen as the others did. But they, like the others, have picked their own branch. Excitement!

When I put my hands on the glass of the hive, the temperature has to be over 96 degrees. It is hot to the touch from the activity. I've never read about INTERIOR temperatures reached by swarm activity, but I know it is 96 degrees or over.

Some observations re: Activity in Hive During Swarm:

1. There is not ONE SINGLE BEE unaffected at those moments. During exit and prior to exit, all interior and exterior bees are in a frenzy.
2. At the time of swarm, all activity stops.
3. ALMOST every bee affects the same body movement (Swarm Dance):
    a. A frantic moving of the whole body back and forth.
    b. Frantic movement of the legs.

c. Aimless moving in place OR aimless moving about in the hive.
4. Shortly AFTER the exit, the hive is quiescent and work goes on as though nothing had happened, but at the MOMENTS, at the NONCE of swarm---all are engulfed.

June 7, 1987

Cloudy, semi-drizzling. 62 degrees.

They are definitely swarming again. But it is an unusual swarm. Tens of thousands of bees are swarming outside and near the bee gate. They are all over the window, all over the hydrangea bush, all over the side of the house. YET there is no swarm CENTER. No swarm on the evergreen. Just all of them out swarming on the house and plants. I just remembered that all day yesterday the queen was piping. I heard her all day as I worked in the kitchen. I'm sure that is the signal for swarming as I recall this same signal before other swarms.

NOW:
1. It is drizzling and they are not leaving the window or bee gate area and thousands are circling overhead.
2. As I look up, I see myriads of bees making crosses in the air. It's fabulous to see them crossing and recrossing in their frenzy.
3. They are in the thousands on the eave of the roof and on the side of the house.
4. After about 30 minutes, they refused to leave the hive area.
5. At 1:40 in the afternoon, they are in a beard on top of the window. The beard tumbles down and

spills over the window, the window sill and the side of the house under the bee gate. What an unusual swarm!

6. The interior of the hive has returned to normal in spite of the close proximity of the swarm. The swarm is, after all, right outside the bee gate!

7. 11:00 at night. The queens are piping like crazy. The swarm of about 3,000 bees (with the old queen there perhaps) is hanging on top of the window between the storm window and the regular window. The rest of the swarm has reentered the hive. I'll see if they leave with this embryo swarm tomorrow or whenever the weather is propitious.

The queens' piping sounds range from c. 7 Yeee to c. 35 Yeee. Some Yeee are muted. Those queens are probably still in their cells. Some sounds are very clear and unimpeded. They have...

I'm back. Just was thinking about the queen's piping to each other. The battle of the queens sort of reminds me of Amazon women warriors. In that early Bronze Age society of women there was only one queen like in my honeybee warrior society. I remember one of the 12 Labors of Hercules was to capture the golden girdle of the Queen of the Amazons, Hippolyta. The ancients definitely believed that there was a tribe of women warriors they called the "Amazons." They were either in Africa or around the Black Sea somewhere. They were formidable foes against men in battle. In order to use the bow better, they cut off their right breasts. They would mate with neighboring tribes in order to continue their female-only society. If the baby was a girl, they would keep her. If a boy was born, they would kill him. In my women-dominated hive, the queens fight to the death as I'm sure the Amazon queens did. All the work, in-hive and out of the hive, is done by the women as with the Amazons.

My bees keep drones around for procreation as did the Amazons. The protection of the hive is solely in the hands of the female bees as was the case with the Amazons. Interesting analogy here. I'm sure it breaks down if you follow it on and on. A big "for instance" is that my honeybees are kind to ALL the young they are raising---girls and boys! But, Sandy, don't forget the Slaughter of the Drones! Retract that?

Note inserted later: Diodorus (c. 340 B.C.), the great Greek historian writes: "...they (Amazon women) thought that the breasts, as they stood out from the body, were no small hindrance in warfare; and in fact it is because they have been deprived of their breasts that they are called by the Greeks Amazons." In the Greek "a-mazos" means "without breast." Interesting.

June 8, 1987

9:30 A.M. Sunny. 80 degrees. The swarm PLUS half of the bees IN the hive are gone---long gone. They seem to be exiting the hive for some unknown, to me, reason.

August 30, 1987

Again I haven't written anything in this book for about 3 months.

All but several hundred bees are dead!! In July the hive functioned well. Up until yesterday the status quo was normal. Today they are dying en masse, in droves, within hours!!!

I feel AWFUL!!

There must be a disease that wipes them out so quickly.

Kent and Jesse (ages 12) came down from upstairs to see this phenomenon of bee death. They feel terrible, too. The children who come in the house have always looked at the bees and have developed an affection for them. Am making Hamburger Mustards on English Muffins for the boys.

August 31, 1987

This group of bees lasted from April 1986 to August 1987. They are the longest-lived of all the batches I've had AND they were, at their peak, the best.

Now it is a holocaust within the hive. Tonight---it is midnight---one lone bee staggers around dazed and semi-paralyzed. She, dying, stumbles over the bodies of thousands of her dead sisters stacked and packed haphazardly at the bottom of the hive.

Here and there a dead bee, suspended by a leg, hangs from the rim of a hexagon.

What could have caused this terrible, swift plague??

They were diminished in numbers since June, but the ones who remained were healthy and active 24 hrs. before they suddenly died.

I'm observing the lone survivor. She staggers around. She tries to crawl up the glass, gets about 3" up and falls back into the pile of dead bees. She tries to walk and stumbles. Her legs seem unable to grip anything for long. She appears to be dragging her abdomen as though it were paralyzed.

I look at her through the magnifying glass. She appears normally formed. I find no outward manifestation of this plague.

I know it has to be some microorganism that has obliterated the bees.

I'll call Ed and see if he has any ideas.

I've looked up Bee Diseases. There is one called Acute Bee Paralysis. Maybe that is it. If so, it is more swift and deadlier than the books indicate. Whatever it is, it killed all but one or two bees in 24 hours!

Here the survivor comes again. She desperately tries to get a foothold. Cling. Cling! She can't. She tumbles and rolls back into the dead.

If she were healthy, she could consume stores and exist. She's definitely slowly succumbing to whatever killed her kin.

Bless her, Lord. Bless them all. I look forward to the day, The Great Day, when the lion lays down with the lamb and disease and death are swallowed up in Victory.

I commend my little ones to You, the Author AND the Finisher of Life.

She is almost dead now. Lying down among her dead sisters, she awaits her release. So much getting and begetting. So much flowering and foraging. So much death and dying. So much hope and living...

INDEX

Abdomen of honeybee 12, 25, 30, 31, 40, 50, 66, 133, 140, 175, 183, 195, 213, 245, 281, 288
Aborted larvae 42, 266
Abortion 49, 310
Alexander the Great 156, 210
Amazon women 314, 315
Amber, gemstone 114, 115
Antenna(e) of honeybee 50, 138, 159, 199, 212, 238, 283
Ant 60, 109, 155, 189, 211, 240, 241
Aphids 240, 241
Attendants, the queen's 8, 29, 31, 49, 50, 103, 183, 184
   Assist her in the laying of eggs 184

Beard of bees 36, 54, 64, 75, 85, 289, 313
Bee diseases 141, 142, 143, 182, 186, 316, 317
Beehive Cluster (of stars) aka Praesepe 18, 48, 49
Beeline 5, 99
Beeswax 12, 15, 23, 29, 100, 132, 153, 154, 170, 175, 191, 200, 210, 213, 255, 258, 282, 305, 306
   Laying of (making) 12, 16, 17, 19, 71, 79, 113, 200, 209, 281
   Uses of 282
Bee gate 7, 18, 43, 53, 62, 98, 138, 153, 197, 311
Bee venom therapy 66
Beowulf 259-262
Bernoulli, Jacob 298
Blake, William 19, 95, 102, 252, 281
Brood 23, 24, 29, 38, 41, 43, 46, 60, 61, 63, 76, 77, 90, 105, 124, 151, 158, 164, 186, 194, 195, 250, 255, 267
Brood comb 8, 11, 13, 26, 30, 36, 48, 64, 91, 93, 103, 123, 135, 159, 175, 210, 212, 214, 287, 305
Brown, Blake 1, 4, 6, 8, 9, 26, 32, 46, 52, 89, 91, 92, 93, 104, 105, 106, 110, 125, 138, 158, 159, 168, 229, 246, 252, 280, 287
Brown, Kathy 1, 37, 52, 53, 56, 59, 89, 99, 105, 138, 155, 213, 229, 234, 246, 252, 254, 280, 287
Browning, Elizabeth Barrett 291
Bubonic Plague aka Black Death 170-172
Bumblebee 44, 104, 117, 215, 270
   Mating of 91-93, 274, 275
Bunyan, John 169

Buzz, buzzing 7, 39, 57, 65, 66, 79, 91, 203, 237, 270, 273, 280, 289
Canning, preserving of food 192, 193
Centrist tendency of honeybee 127, 128, 131, 214, 247, 267
Cleaning of hive by honeybees 53, 64, 80, 128, 132, 139, 225, 236, 283
Coleridge, Samuel Taylor 168, 169
Cross-pollination 96, 193, 194, 215, 233, 234
Cryonics 147-149

Daedalus (and Icarus) 153, 154, 155
Dances of bees, in-hive and directional 61-63, 80, 133, 162, 203, 204
  Crescent 204
  Embracing 133, 137, 138
  Group 160, 162, 204
  Pollen 61, 62, 63, 101, 108, 204
  Pull 204
  Round 204
  Shaking 61, 133, 256
  Swarm 312
  Wag-Tail 204
Dore, Gustav 168
Drones 13, 14, 22, 26, 33, 34, 35, 43, 62, 66, 86, 87, 119-121, 122, 123, 164, 176, 243, 250, 275, 276, 277, 286, 315
Dwindling in winter 118, 127, 150, 158, 170, 183, 248, 302

Eggs of honeybee 16, 30, 31, 45, 50, 249
  Fertilized 35, 45, 87, 274, 275
  Unfertilized 164, 275, 276
Etymology 14, 15, 96, 257
Eunuch 122, 123
Eyes of honeybee 22, 87, 94, 95, 176, 177

Fanners and fanning, honeybee 51, 68, 69, 98, 99, 107, 113, 141, 145, 238
Feelers of honeybee 132, 183, 184, 219
Fibonacci aka Leonardo da Pisa 116, 117, 231, 232, 300
Film noir 129, 130
Foragers, foraging 23, 44, 48, 98, 100, 119, 139, 164, 178, 218, 221, 248
Formations, military 23, 227, 236, 238

Gordian knot 156
Grooming, honeybee 29, 112, 180, 183, 258

Guard bees 55, 56, 76, 98, 118, 120, 222, 227, 228, 230, 231, 237, 238, 239, 241, 242
Hexagons 12, 16, 29, 30, 41, 79, 95, 127, 128, 132, 255, 256, 260, 267, 306
Heyerdahl, Thor 114, 201
Hive(s)
  Bee gums 7, 78
  Nucleus aka "nuc" hive 202, 203
  Observation. See under Observation Hive
  Outside hives aka supers 7, 9, 54, 160, 225
  Paris Opera House hive 247, 248
  Skeps 18, 78
  Tile 78
Hiving of honeybees 5, 6-11, 195, 196, 251, 280, 303
Hive tool 7, 199, 251, 278
Honey 5, 15, 26, 27, 34, 38, 40, 45, 71, 86, 94, 100, 124, 127, 128, 133, 136, 158, 161, 177-179, 188, 189, 190, 191, 193, 206, 208, 210, 216, 229, 235, 236, 248, 259, 260, 268, 278, 291, 293, 304
  Consumption of in winter cluster 124, 127, 135, 136, 190, 191, 308
  Fermentation of 191, 259, 260, 261, 276, 277
Honeybee, Apis mellifera 23, 40, 92, 153
  Birth of 45, 46, 169, 195
  Carniolan 288, 294, 302
  Italian 37, 198, 288
  Life span of 36, 90, 173
  Population of in the hive 6, 7, 10, 20, 60, 76, 203, 285, 307
Honeycomb 12, 16, 47, 79, 91, 123, 128, 139, 174, 178, 190, 199, 223, 255, 256, 277, 305
Hornet's nest 104
Hum, humming 11, 43, 48, 49, 83, 144, 162, 211, 265, 269, 270
Hummingbird 97
Hummingbird moth 270
Hurricane Gloria 295

Julius Caesar 10, 165, 166, 192

Killer bees 4, 207, 208, 252

Langstroth, Rev. L.L. 238, 253, 259
  <u>Langstroth On The Hive And The Honeybee. A Beekeeper;s Manual</u> 238

Larva(e) 35, 39, 40, 41, 42, 48, 50, 98, 164, 175, 220, 222, 224, 263, 266, 277, 303, 312

Martial, Marcus V. 114, 115
Mating of honeybees 86, 87, 88, 92, 119-121
Maya 122, 129, 201, 271, 272
Mead, alcoholic drink made from honey 38, 188, 259, 260, 261, 290, 291
Milne, A.A. 137
Minoan 89

Nautilus, spiral sea shell 155, 297, 298, 300
Necklace (of bees) 16, 17, 21, 209, 281
Nectar 7, 15, 18, 23, 27, 29, 38, 51, 67, 77, 94, 95, 96, 97, 99, 100, 111, 115, 159, 177, 179, 193, 216, 229, 230, 242, 259, 260, 278, 286, 304
Nectar crop of honeybee 27, 119, 283, 286
Nurse bees 35, 98

Observation hive 4, 5, 6, 13, 91, 120, 229, 256
   Design problems of 91, 113, 140, 143, 150, 180, 181, 184, 199, 200, 217, 244, 285
Oswego tea 215, 270

Pan, the Roman satyr 290-292
Parades, in-hive
  Of the dead 21, 22, 222, 224, 244, 246
  Of larva 220, 221, 224, 244, 266
  Of intruders 222, 224
Parthenogenesis (and drone-laying workers) 164, 276, 305
Pheromones (smell) 36, 65, 87, 120, 163, 164, 165, 176, 229
Pollen 13, 15, 22, 23, 29, 35, 38, 47, 67, 98, 100, 101, 102, 108, 111, 124, 125, 126, 193, 204, 217, 230, 238, 242, 302, 305
Pollen basket of honeybee 22, 24, 53, 59, 61, 62, 117
Pompeii 192, 193
Proboscis aka nectar tongue 21, 25, 45, 61, 65, 83, 112, 133, 138, 176, 179, 180, 183, 184, 195, 242
Propolis aka bee glue 20, 97, 99, 100, 105, 141, 152, 159, 160, 175, 187, 194, 199, 209, 210, 211, 218, 225, 228, 234, 235, 238, 259, 265, 296

Queen bee 7, 8, 9, 10, 11, 14, 21, 29, 31, 35, 40, 49, 55, 66, 74, 85-87, 93, 102, 119-121, 164, 184, 287, 288, 289, 312, 314

Balling of 36
Candy plug of at hiving 13, 252, 304
Clipped 34, 35, 54, 55, 63, 65, 67, 91, 267, 277, 289
Laying habits of 29, 30, 31, 32, 36, 45, 46, 51, 103
Piping of 73, 74, 80, 81, 83, 84, 289, 292, 313, 314
Queen cell aka supercedure cell aka gyne cell 34, 35, 36, 41, 51, 63, 74, 93, 269, 292
Queen excluder 9, 34, 121, 250, 287, 288, 289

Royal jelly 35

Scout bees 74, 76, 77
Shakespeare, William 134, 166, 167, 249, 290
Shroud of Turin 124-126, 201
Silk Road 27, 28, 303
Silk worm 27, 302, 303
Silver, Jesse 1, 6, 8, 9, 42, 52, 53, 56, 65, 66, 89, 104, 105, 106, 123, 125, 137, 138, 141-144, 154, 155, 168, 195, 219, 229, 246, 280, 295, 316
Silver, Steve 1, 4, 5, 7-11, 34, 52, 70, 89, 93, 104, 144, 155, 187, 196, 197, 200, 202, 210, 246, 251, 252, 260, 272, 280, 295
Slaughter of the drones 119-121, 243, 315
Spooner, Rev. William 253, 254
Sting, bee 36, 55, 65-67, 74, 90, 164, 206, 207, 223
Stinger of honeybee 33, 66, 164, 165, 206, 251, 277
Sun, importance of to honeybee 12, 19, 24, 40, 48, 50, 63, 67, 68, 79, 84, 112, 139, 164, 189, 273, 295, 296, 307, 308, 311
Suspended animation 145-149
Swarm, swarming 36, 52-55, 59, 60, 63, 67, 68, 75-79, 84-86, 225, 267, 268, 269, 288, 289, 292, 293, 294, 312-314
    Aborted swarm 54, 67, 143
    Storm swarm 88, 281
    Sweet swarm 90, 294

Tasks (in-hive) 53, 80, 98, 99
Twins
    Larvae 263, 266, 276
    Minnesota Twin Study 264

Unanswered questions 19, 30, 31, 33, 35, 40, 43, 46, 48, 50, 59, 60, 64, 76, 77, 79, 83, 86, 99, 105, 107, 108, 132, 133, 138, 174, 184, 199, 216, 227, 244, 275, 307, 317

Ventris, Michael 89

Wax moth 276, 277, 279, 302, 303, 304
Williams, Kit 297, 299, 300
Wings of honeybee 45, 51, 83, 87, 131, 139, 153, 238, 290
Winter cluster 120, 130, 135, 300, 309
Wordsworth, William 291
Worker bees 15, 33, 35, 36, 39, 43, 46, 54, 62, 74, 98, 119, 121, 132, 164, 198, 200, 203, 204, 214, 238, 256, 275, 276, 277, 281, 286

Xenophon 178

Yellow jacket bees 215, 221, 224

Made in the USA
Coppell, TX
09 October 2021